T0130761

THE LOGIC OF IMMUNITY

The Logic of Immunity

•

Deciphering an Enigma

B. J. CHERAYIL

JOHNS HOPKINS UNIVERSITY PRESS | *Baltimore*

© 2024 Johns Hopkins University Press
All rights reserved. Published 2024
Printed in the United States of America on acid-free paper
9 8 7 6 5 4 3 2 1

Johns Hopkins University Press
2715 North Charles Street
Baltimore, Maryland 21218
www.press.jhu.edu

Library of Congress Cataloging-in-Publication Data is available.

A catalog record for this book is available from the British Library.

ISBN 978-1-4214-4765-0 (hardcover)
ISBN 978-1-4214-4767-4 (ebook)

Special discounts are available for bulk purchases of this book. For more information, please contact Special Sales at specialsales@jh.edu.

For Nandini, Monisha, Maya, Soraya, and Amachi

and

To the memory of my father, Prof. Joseph D. Cherayil

CONTENTS

Preface   *ix*

1. Conceptualizing the Immune Response   1
2. Innate Immunity I: The Rapid Response Team   10
3. Innate Immunity II: Sensing Infection   24
4. Adaptive Immunity: Strength in Diversity   35
5. B Lymphocytes: Antibodies "R" Us   49
6. T Lymphocytes: A Little Help for My Friends   66
7. Immune Dysfunction: When Things Go Wrong   85
8. Conditioning of the Immune System by the Microbiota   104
9. Remembrances of Things Past Haunt the Present   120
10. Harnessing the Function of the Immune System   131
11. Vaccination: The Power of Prevention   148
12. "To Follow Knowledge Like a Sinking Star"   160

Acknowledgments   165
Abbreviations   167
Glossary   169
References   181
Index   185

I wrote this book with the goal of providing a concise, lucid introduction to the immune system, the consortium of cells and molecules that work together to keep us safe from infection. It is based on the many years that I have spent teaching immunology to undergraduates, medical students, physicians, and biomedical researchers. I was motivated to convert that accumulated experience into a manuscript by the recent pandemic, a global calamity that has suddenly made antibodies and T cells topics of general interest.

I have focused on conveying the basic principles that govern the functioning of the immune system and have kept the finer details to the minimum needed for clarity and for an appreciation of the underlying mechanisms. I have also tried to make the explanations accessible to as wide an audience as possible. It is difficult, however, to discuss a subject like immunology without using a few technical terms that may be unfamiliar to the general reader. With that in mind, I have provided a list of commonly used abbreviations as well as a glossary of the relevant nomenclature at the end of the main text.

Overall, I would like to think that the book will serve as an easily understandable synopsis of the key concepts, a synthesis that is comprehensive but not dauntingly complex. So, if you have ever wondered how exactly the immune system fights off the pathogens that may assail us or why it sometimes malfunctions and causes disease, I hope that the information in the pages that follow will help to answer those questions. I also hope that it will tempt you to look more deeply into a subject that is both endlessly fascinating and directly relevant to our health and well-being.

THE LOGIC OF IMMUNITY

CHAPTER 1

Conceptualizing the Immune Response

Imagine for a minute that you are the commander of an expedition to a remote planet, one that holds promise as a refuge from the deteriorating conditions on Earth. As you orbit this new world and prepare for landing, you receive word that a group of extraterrestrials, who are also fleeing a dying home, has already taken up residence a short distance from the area where your spaceship is about to touch down. You do not know whether they are friendly or hostile, but you have been instructed by mission control to prepare for any eventuality. What would you do? How would you ensure the safety and survival of the people under your care?

One of your first actions as you set up camp will probably be to order the construction of a perimeter fence, a barrier that will enclose your colony and provide initial protection against trespassers. Since the fence can act only as a deterrent and cannot be made completely impervious to intrusion, it would be prudent to post sentries at strategic points within the perimeter so that they can take action if the barrier is breached and an unauthorized individual gains entry. Of course, you will have to keep in mind the possibility that the sentries might be overcome if hostile forces have superior numbers or more sophisticated weaponry. It would be a good idea, then, to equip your guards with the means to sound an alarm, so that if it becomes necessary, they can call in mobile backup forces that can help to repel the invaders. In such circumstances, they should also be able to send a message to camp headquarters to alert leadership of impending danger, giving those in charge sufficient time to organize and mount a defense that is most appropriate for the type of intruder involved.

This imaginary scenario illustrates the general organization and functioning of our immune system, the collection of cells and molecules that protects most multicellular organisms, including humans, from infection by microorganisms. Each of the defensive measures used by our expedition of intrepid earthlings has an immunological equivalent (figure 1.1). The perimeter fence is the tightly packed layer of cells—the epithelium—that covers our skin and lines the surfaces of our gastrointestinal, respiratory, and genitourinary tracts. The epithelium constitutes the frontier that separates the microbe-packed external world from the relatively sterile environment of our body's interior (although the cavities or lumens of our gastrointestinal, respiratory, and genitourinary systems are generally viewed as being inside our

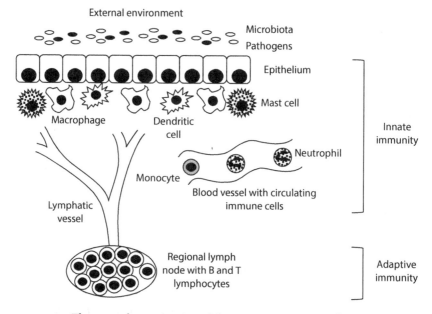

FIGURE 1.1. The general organization of the immune system, as illustrated by the main cellular elements involved in the immune response in a representative peripheral tissue.

bodies, they are in communication with, and so are considered to be a part of, the exterior). The sentries are the frontline immune cells—macrophages, dendritic cells, mast cells, and others—that reside in every tissue, including just beneath the epithelium, and that have the ability to recognize microbial invaders and respond within a matter of minutes. The functions that they deploy include engulfment and killing of microorganisms, and the release of a variety of molecules that signal to the roving patrol of backup defenders—circulating immune cells such as neutrophils and monocytes—that they should exit the blood vessels and move into the tissue to join the fight.

During this early phase of the response to microbial intrusion, dendritic cells carrying fragments of the microorganisms that they have ingested sneak out of the tissue through thread-like lymphatic vessels and migrate to the regional lymph node, functioning like dispatch riders conveying messages from the battlefield to the local command post. And, indeed, lymph nodes—discrete, organized collections of B and T lymphocytes that are located at strategic sites around the body—are where higher-level immunological decisions are made. The arrival of the microbe-laden dendritic cells alerts the B and T cells that there is a potentially dangerous situation developing in the tissue region that is served by that particular node, and that a response is required. The process of mounting the appropriate lymphocyte response takes time, usually several days. During this period a team of B and T cells, adequate in number and with capabilities most suited to dealing with the microorganism that is the cause of the problem, is assembled and equipped. The two types of lymphocytes have specialized and complementary functions. The B cells secrete antibodies, which are soluble proteins that circulate through the blood and specifically target the microbes for elimination. Meanwhile, the T cells migrate out of the lymph node and travel via the circulation to the affected tissue, where they work cooperatively with each other and with the local immune cells to clear the infection.

I first became interested in the immune system soon after graduating from medical school in India, when I was assigned to an internship rotation in the community health department of the affiliated hospital. One of my regular responsibilities there was to help staff a mobile clinic that served several outlying villages. Another intern and I, along with a couple of nurses and a pharmacist, would depart early in the morning in an ungainly, blunt-nosed vehicle, essentially a giant box on wheels, which had been acquired by the Indian government as part of a deal with a British company and that had somehow found its way to our institution. Contained within its commodious interior were an examination cubicle, a small surgical suite, and multiple shelves, cupboards, and fold-out tables that were ingeniously tucked into the walls. It was a marvel of modern technology, designed to take state-of-the-art medical care to even the most remote regions of low- and middle-income countries. Unfortunately, it was completely unsuited to the driving conditions in our area, particularly the narrow, winding, and largely unpaved tracks that snaked from one village to the next. Even on the marginally better roads closer to town, riding in it was torture. Unless you were lucky enough to snag the well-cushioned seat next to the driver, you were forced to sit in the back and endure a hot, excruciatingly uncomfortable ride, constantly jolted from side to side or bumped up and down as the lumbering vehicle, which we had christened the White Elephant, made its slow, ponderous way to its destination.

At each village on the route, we set up a folding table under the shade of the nearest tree and tended to the twenty or thirty people who were waiting to be seen. Most of them had relatively minor problems that could be taken care of with the basic medicines that we brought with us, and that the pharmacist dispensed through a small window in the side of the truck. There were usually a few pregnant women who needed antenatal care and, on occasion, someone had an injury that required dressing or stitching up. Anything more serious had to be sent to the main hospital. Many of the villages had several patients with leprosy, who relied on the clinic for their monthly supply of dapsone.

Leprosy, or Hansen's disease as it is more formally called, is an illness that is usually associated with ancient times, something you might encounter in the Bible or in the dark ages, but it was very much an active problem in India in the 1980s. It was a terrible disease—disfiguring, incapacitating, and requiring years of treatment with the drugs then available. Those who were unlucky enough to contract it were often ostracized and forced into a life of begging on the streets. But it was also an intriguing disease. Caused by the slow-growing bacterial pathogen *Mycobacterium leprae*, it manifested with clinical abnormalities that were distributed along a spectrum, with the two extreme or polar forms having very different characteristics. At one end of the spectrum, patients with lepromatous leprosy were unable to control the growth of the bacteria, and their skin became distorted and lumpy with the accumulation of large numbers of the organisms. As the infection progressed, nerves, bones, and other tissues were gradually affected, leading in the late stages to loss of sensation, destruction of the nose and fingers, and several other complications. At the other end of the spectrum was tuberculoid leprosy. In this form few, if any, *M. leprae* were found in the tissues but impairment of nerve function occurred early in the disease, presenting as discrete but often hard to detect areas of numbness and subtle discoloration on the skin or as weakening of the muscles of the hand and foot. If left untreated, the nerve damage could become severe and could result in grossly deformed and barely functioning limbs. Between the two ends of the leprosy spectrum were several intermediate varieties with features that were hard to classify or that represented a mix of the polar forms. Finally, there was one additional clinical category, which was represented by the people—health care providers, social workers, family members of patients—who, despite repeated and sometimes prolonged contact with individuals teeming with *M. leprae*, never developed any form of leprosy.

Same pathogen, very different outcomes. What was the explanation for the striking divergence of clinical course and pathology between the lepromatous and tuberculoid forms of leprosy, and for the fact that most people who were exposed to *M. leprae* did not manifest with the disease at all? That question does not apply uniquely to leprosy, of course. Other infections also display a similar heterogeneity of clinical outcome. Active tuberculosis (TB) develops only in a small proportion of individuals who are exposed to or infected with the causative agent, *Mycobacterium tuberculosis*. The majority of people either clear the pathogen completely soon after infection or successfully contain it without any abnormalities other than a positive skin test and, perhaps, some shadows on a chest X-ray. Severe acute respiratory syndrome coronavirus 2 (SARS-CoV-2) causes a mild flu-like illness in most individuals, but it can put 10% of those infected in an intensive care unit with life-threatening problems. Even the ubiquitous rotavirus induces diarrhea in less than half of the children that it infects.

The issue of clinical heterogeneity can be extended to several noncommunicable diseases too. Why is it, for instance, that most people can eat peanuts or shellfish to their heart's content, whereas a few unfortunate individuals break out in hives, struggle to breathe, or even go into shock if they get so much as a whiff of a nut or a shrimp? Why do some people develop inflammatory bowel disease (IBD)? Or systemic lupus erythematosus (SLE)? These questions were not easily answered at the time that I was in medical school. There was a general suspicion that abnormalities of the immune response were involved, but what they were and what role they played no one knew.

The immune system has been the focus of scientific study for over a hundred years. But it was probably only in 2020, when SARS-CoV-2 was raging around the globe, and terms like "cytokine storm," "neutralizing antibody," and "herd immunity" acquired immediate urgency, that the functioning of our macrophages and lymphocytes became of interest to people outside of the laboratory. Suddenly, immunology was the hot topic on the news and on social media. Immunologists were taking to Twitter and TikTok to counter misinformation and swirling conspiracy theories, while one of the most prominent representatives of the field was being vilified or applauded depending on the political loyalties of the commentator. All the attention seemed appropriate given the make-or-break role played by the immune system in determining the outcome of SARS-CoV-2 infection: a properly functioning immune response could stop the virus in its tracks and save your life, especially with a little help from vaccines; if the response went awry, however, it could wreak more havoc on your body than the infection itself.

But the immune system is more than just the nemesis or dupe of pandemics. It is what allows us to coexist in relative harmony with the huge numbers of microorganisms that are *always* around us, including those that reside in intimate association with our skin and the linings—the mucosa—of our gastrointestinal, respiratory, and genitourinary tracts. The trillions of bacteria, viruses, and fungi that make up our resident microbial community, known as the microbiota, are harmless for the most part, and many actually do us a lot of good. But some of these microbes, as well as any actively pathogenic organisms that might be present in our more general environment, could invade our tissues to feast on the abundant nutrients available there if our immune cells had not figured out over the course of evolution how to deal with them. The mechanisms that have developed for this purpose, and that maintain a delicate balance between preventing uncontrolled microbial intrusion into our bodies and allowing some organisms, including the beneficial ones, to reside on specific tissue surfaces, are as beautiful in their functional artistry as the workings of a fine timepiece or the structure of a medieval cathedral. Understanding these mechanisms is reason enough to embark on a study of immunology. If additional motivation is required, consider that many, perhaps most, human diseases have an immunological component, from the obvious aberrations of immunity in the acquired immunodeficiency syndrome (AIDS) or lupus to the more subtle involvement of the immune system in obesity, atherosclerosis, and Alzheimer's disease. Correcting the abnormalities of immune function in these conditions has the potential to attenuate the pathologic process, relieve symptoms, and even bring about complete cure. Such immune-based therapeutic strategies have already met with notable, sometimes spectacular, success in previously intractable problems like IBD, rheumatoid arthritis, and certain types of cancer. And, of course, vaccination against a number of infectious diseases, including coronavirus disease 2019 (COVID-19), the illness caused by SARS-CoV-2, has saved millions of lives all over the world and probably represents the greatest triumph of immunology. These achievements, and similar ones that will hopefully occur in the future, would not be possible without insights into how the immune system works. That these insights are an ongoing preoccupation of immunologists goes without saying, but they really should be of interest to anyone who cares about health and disease.

Unfortunately, making sense of the intricacies of the immune response is often perceived as a task that is too daunting to undertake. The multiple cell types involved, the array of molecules that each cell type expresses, and the sometimes abstruse nomenclature all conspire to make immunology needlessly bewildering and probably discourage people from giving the subject more than a passing glance. Beneath the complexity, however, are some basic concepts that explain and illuminate the key aspects of immune function. Once these ideas are understood, the rest of immunology is essentially a matter of detail.

When I teach introductory immunology, I usually employ the analogy of the space expedition or something similar to provide the framework for thinking about a typical immune response, the essential elements that are needed to mount an effective defense against intruders. It serves to introduce the key cellular players as well as the two main stages of the response (see figure 1.1): innate immunity, which acts very rapidly and broadly and is mediated largely by epithelial cells and specific immune cells that are resident in tissues or recruited from the circulation; and adaptive immunity, which takes longer to develop, is more specialized and flexible in its action, and is the function carried out by T and B lymphocytes. I invite you to join me in examining each of these aspects of the immune system in greater detail in the chapters that follow, starting with the innate immune response, then going on to an overview of adaptive immunity, followed by a discussion of the specific roles played by B cells and T cells. By the end of chapter 6, I hope that the logic, perhaps even the elegance, of the cellular and molecular interactions that are involved in the immune response will become clear. With this information in hand, we will go on to discuss how immune functions can fail or misfire because of genetic and nongenetic influences, giving rise to diseases characterized by deficient, excessive, or inappropriate immunological activity. Finally, we will consider how the immune system can be controlled and harnessed for the purpose of alleviating clinical problems. Perhaps, through the course of these discussions, we will get some insight into why a few individuals exposed to *M. leprae* develop lepromatous leprosy while others get the tuberculoid form, or why some people develop lupus, IBD, or food allergies.

CHAPTER 2

Innate Immunity I

The Rapid Response Team

"Good fences make good neighbors." When Robert Frost immortalized that bit of folk wisdom, it's unlikely that he was referring to epithelial surfaces. But the dictum applies as well to the cellular layer that is our interface with the exterior as to a barricade between adjoining properties. The purpose of both structures is to demarcate territory, prevent unwanted intrusions, and maintain peace between communities that have to live in close proximity. In the case of our bodies, the barrier tissues that are immediately adjacent to the external world—the skin, the upper and lower portions of the gastrointestinal tract, the upper part of the respiratory tract, and the lower genitourinary tract—are constantly exposed to the large variety of microorganisms that are ever-present in our environment. Among the teeming millions of microbes that can be found in the air that we breathe, in the food and drink that we ingest, and on the objects and people that we come in contact with, there may be some that have evolved mechanisms to interact with our tissues in a way that results in serious cell damage. Such organisms are described as pathogens because the tissue abnormalities—the pathology—that they cause can lead to the development of disease.

In addition to the microorganisms present in the general environment, barrier tissues are exposed to a large and diverse collection of bacteria, viruses, and fungi that live in the lumens of our gastrointestinal, respiratory, and genitourinary tracts and that colonize the surface epithelium of these structures as well as the skin. Together, these resident organisms make up our microbiota, a thriving microscopic community that represents our closest microbial neighbors. For those interested in numbers, it has been estimated that there are 38 trillion bacteria in this community, a little more than the total population of human cells in the body. If we add the myriad fungi and viruses that are also constituents of the microbiota, we arrive at the somewhat disturbing realization that we are all "meta-organisms," with more than 50% of our composition being microbial in nature.

Fortunately, most of our resident microorganisms are harmless under ordinary circumstances and many are actively beneficial. For example, some of the bacteria that live in our intestine help in the digestion of dietary constituents like fiber that we would otherwise not be able to utilize, while others synthesize vitamin K and members of the vitamin B family that we cannot make ourselves. And yet, along with the "good" and "not evil" microbes, the microbiota may also harbor some potential bad actors—organisms with latent virulence characteristics that could allow them to become opportunistic pathogens if the conditions were right (or wrong, from our perspective). If such organisms were to grow uncontrolled and unimpeded, they could cause considerable harm. That this can actually occur is apparent from the observation that shortly after we die some of the bacteria that are normally confined to our gut or skin spread beyond those sites and become detectable in large numbers in the blood and deep tissues.

The Wall

So, what is it that normally protects us from pathogens in the environment and from potentially dangerous members of the microbiota? The first line of defense is the epithelium, which is the tissue that covers the surfaces of our skin and mucosa. The epithelium can be made of many layers of cells, as, for example, in the skin, mouth, esophagus, or bladder, or it can be just a single layer thick, as in most of the intestine. In either case, the cells are linked to each other by intricate assemblies of junctional proteins, forming a continuous mechanical barrier that resists penetration by most microorganisms. But the epithelium is not just a passive border wall. Rather, its constituent cells actively contribute to its protective function (figure 2.1). Some are specialized to produce a surface coating that reinforces the mechanical properties of the epithelium. In the gastrointestinal, respiratory, and genitourinary tracts this coat is a slimy layer of mucus proteins; in the skin it is made of oily, water-repelling lipids that create a tight seal around the outermost cells. Other cells of the epithelium secrete antimicrobial peptides, small molecules that ensnare bacteria in a dense, incapacitating mesh or kill them outright by poking holes in their membranes. Such cells have a characteristic appearance in the intestine and are known as Paneth cells. In addition, a handful of epithelial cells—the stem cells—are endowed with the ability to generate all the others, either continuously or when needed, ensuring the timely repair of minor defects in the barrier that might result from attrition or damage.

Many of the epithelial functions that help to prevent microbial intrusion—the expression of mucus proteins and antimicrobial peptides is a good example—are actually induced by the microbiota. Our resident microbial community thus equips the epithelium with the

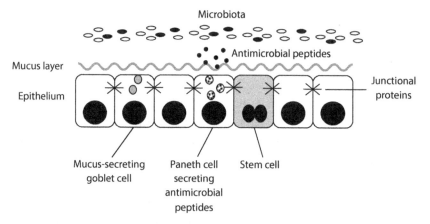

FIGURE 2.1. Protective properties of the intestinal epithelium. Similar features are present in the epithelium of other barrier tissues, such as the skin, respiratory tract, and genitourinary tract.

very properties needed to control potentially virulent microorganisms, including bad elements of the microbiota, a circumstance that turns the Frost aphorism on its head: good neighbors make good fences. This is a recurring theme in immunology. Several aspects of both innate and adaptive immunity are programmed by exposure of immune cells to various molecules expressed by the microbiota, giving our microbial neighbors a major role in determining immune function. We will learn more about the intimate but uneasy relationship between the microbiota and the immune system in chapter 8.

The epithelium helps to enforce a truce that has been established between adjoining cellular and microbial communities over the course of evolution. Maintaining this state of peaceful coexistence allows both parties to the agreement to reap significant rewards. Our resident microbes get to live in an environment that they have adapted to and that provides several nutrients needed for their survival and growth. We (as represented by our cells) avoid the tissue damage that would inevitably result from either invasive growth of microorganisms or from our response to them, and also benefit from several functions carried out by the microbiota: digestion of nutrients, generation of useful metabolites, including certain vitamins, and protection from the incursions of environmental pathogens.

The last function—protection from pathogens—depends in part on microbiota-mediated programming of the immune system, but it also reflects additional mechanisms by which the microbiota helps to reinforce the barrier properties of the epithelium: competing with incoming pathogens for resources and space in specific tissue niches or producing metabolites that inhibit pathogen growth. The importance of the microbiota in resisting pathogen colonization becomes obvious when it is depleted by prolonged antibiotic treatment, for instance, in individuals who have chronic infections or who are on immunosuppressants following a transplant. Like a vacant building becoming prey to squatters, the gastrointestinal tract that lacks its normal microbial residents becomes susceptible to several pathogens, most notoriously to a bacterium called *Clostridium (Clostridioides) difficile*, which illicitly occupies the colon and causes a severe and very treatment-resistant form of diarrhea.

Defenders in Residence

Although the epithelium is a robust, self-repairing barrier, it can be weakened by mechanical, chemical, thermal, or other types of injury and allow the entry of pathogens or potentially virulent members of the microbiota that are normally kept in check. Significant breaches of the epithelium, whether they are secondary to weakened barrier functions or the result of active intrusion by pathogens, are handled by a second line of defense consisting of the immune cells that reside in almost all tissues, including in the vicinity of epithelial surfaces. There are many types of resident immune cells—macrophages, dendritic cells, and mast cells, to name a few—and new ones are still being discovered. Even cells that are not traditionally considered to be part of the immune system, such as the endothelial cells that line blood vessels, nerve cells, and the fibroblasts that contribute to the general background "stuff" or stroma of most tissues, can be co-opted to carry out immunological functions when needed. All these cells may become involved in the immune response in one way or another depending on the circumstances, and we will encounter several of them in later sections. But from the perspective of dealing with microorganisms that have made their way through the epithelium, one of the most important cell types is the macrophage.

As its name indicates, the macrophage is the big eater of the immune system. This amoeba-like cell is singularly skilled at ingesting (or phagocytosing, to use the more scientific term) any particulate material—dead cells, cellular debris, foreign bodies, microorganisms— that it encounters as it crawls slowly through tissue spaces. During phagocytosis, the macrophage extends protrusions of the plasma membrane that engulf the particle being targeted, ultimately enclosing it in a membrane-bound compartment or organelle called the phagosome. The phagosomal cargo is then subjected to a barrage of caustic chemicals, including highly reactive radicals of oxygen and nitrogen, before being delivered to another organelle—an acidic sub-cellular compartment called the lysosome—where its destruction is completed as a result of exposure to degradative enzymes. There are very few microorganisms that can survive this harsh treatment. Consequently, tissue macrophages are efficient defenders of their territory, rapidly eliminating any microbes that might make their way through the epithelium. If the number and virulence of the microbial intruders is low, they are cleared by macrophages with little or no fuss and we are often unaware that a problem has occurred. In fact, it is likely that such trivial skirmishes between intruders and defenders occur several times a day in our skin and other barrier tissues because of minor breaks in the epithelium. That they do so silently, without becoming apparent as a clinical abnormality, is a testament to the effectiveness of macrophage antimicrobial functions. It is only when these functions are missing or defective do we realize how important they are. Consider, for example, the plight of individuals with an inherited condition known as chronic granulomatous disease (CGD), which is caused by a genetic mutation that renders macrophages and other phagocytic cells incapable of generating the reactive oxygen radicals necessary to kill microorganisms that they have engulfed. Patients with CGD are plagued by recurrent infections of the lungs and of the soft tissues beneath the skin, often involving organisms that are not particularly virulent and that the rest of us are able to shrug off thanks to the quiet efficiency of our phagocytic cells.

The person who is widely credited with recognizing the importance of macrophages and phagocytosis in antimicrobial defense is the nineteenth-century Ukrainian zoologist Élie Metchnikoff. He appears in most photographs as a grandfatherly figure, with kindly eyes, a hint of a smile, and the unkempt hair and bushy beard that marked serious men of that period. But by his own account his early life was spent under the shadow of recurrent episodes of depression. He attempted suicide twice: the first time, by morphine overdose, following the death of his beloved wife and the second, by auto-infection with pathogenic bacteria, in response to work-related stress. Yet, he was deeply passionate about his research. It is clear from his writings, especially the description of his famous experiment on phagocytosis in starfish larvae, that he took a child-like delight in his work. The joy that he found in observing and analyzing macrophages under the microscope may have acted as a counterweight to the darker side of his personality, allowing him to generate the data that laid the foundation for a whole branch of immunology and that led ultimately to a Nobel Prize.

Calling in Backup: Distress Signals

Sometimes, tissue macrophages are overwhelmed by the microorganisms that they encounter. This situation may arise either because the microbes are present in large numbers or because they have special virulence characteristics that allow them to resist the killing mechanisms that macrophages deploy against them. All is not lost, however: at the same time that these mechanisms are activated, the macrophage also sends alarm signals to surrounding cells to alert them of impending danger and to recruit backup forces from the circulation. The signals are in the form of secreted molecules that are released by macrophages as well as other cells in the infected tissue and that diffuse outwards from the point of disturbance.

These molecules are of three major types. They include messenger proteins known as cytokines, important mediators of cell-cell communication in the immune system; chemokines, which are cytokine-like proteins specialized for guiding and directing cell movement; and eicosanoids, a diverse group of small bioactive lipid molecules that carry out a number of different functions depending on their identity. Cells in the vicinity perceive these various signals by means of specific surface receptors and change their properties or behavior as a result, with the exact effects being determined by the type of cell and specific cytokine, chemokine, or eicosanoid involved. Most cells will respond to the alarm signals by switching to an activated state in which the expression of multiple defensive molecules is increased, a reaction that provides the means to ward off microbial attack if the need arises.

The classic example of this type of anticipatory defense mechanism is the response to a particular family of cytokines known as type 1 interferons (sometimes also called innate interferons). Type 1 interferons are typically secreted by cells that are infected by viruses, and they act on adjacent, noninfected cells to increase the expression of hundreds of genes that inhibit viral entry and multiplication. The elevated levels of these interferon-stimulated genes enhance resistance to viral infection, thereby helping to limit the spread of the virus and the cell damage that it causes. Early production of type 1 interferons is an important defense against SARS-CoV-2 infection, and individuals who have an impairment in this response often go on to develop more severe disease.

Beyond the type 1 interferons, there are a few dozen additional cytokines that can be released by macrophages and other cells in response to the presence of microorganisms, with the types and quantities being influenced to some extent by the nature of the microbial threat. They have different patterns of expression and function, although there are significant overlaps in many cases. Examples of cytokines that are frequently secreted during the response to infection, particularly in the context of infection with bacteria, are tumor necrosis factor α (TNFα), interleukin-1β (IL-1β), and IL-6. In general, these molecules act on neighboring cells to enhance antimicrobial defenses, promote activating or regulatory functions, facilitate recruitment of circulating cells, or induce the production of additional mediators. But excessive and uncontrolled production of some cytokines, including TNFα and IL-6, can lead to damaging effects on vital organs and, in the worst cases, can result in cardiovascular collapse. The cytokine storm that was a notorious (albeit unusual) consequence of SARS-CoV-2 infection during the early days of the COVID-19 pandemic is an example of this type of unregulated and dangerous immune response.

A large number of chemokines can also be produced by cells in infected tissue. Their nomenclature and functions are complex, and even professional immunologists have trouble keeping track, but the important thing to remember is that different chemokines induce the migration of different types of immune cells from the circulation and through tissues. Some attract neutrophils preferentially, others monocytes or lymphocytes. The precisely regulated production of various chemokines at the right time and location ensures the coordinated recruitment of the appropriate cells to the site of infection.

Finally, the third category of molecule released by cells in response to a microbial encounter, the eicosanoids, includes a variety of small, functionally potent lipids that are broadly classified into leukotrienes and prostaglandins depending on their structure. They share some of the properties of cytokines and chemokines in being able to activate or recruit specific cell types, but they may have additional attributes that help to mediate or control the response to microbial intruders.

Backup Troops to the Rescue

Some of the most important responses induced by the cytokines, chemokines, and eicosanoids that are released during infection occur in the small blood vessels that meander through the affected tissue. The vessels dilate and become leaky, increasing local blood flow and allowing protective plasma proteins to diffuse into the tissue to the site of infection. In addition, the endothelial cells that line the vessels increase expression of specific adhesive molecules, displaying them on their surfaces so that any circulating immune cells that have the right receptors can grab hold. Such receptors are particularly prominent on neutrophils, phagocytic cells that can be easily identified by their segmented nucleus and granular cytoplasm, and that constitute the most potent microbial killing machines of the immune system.

Neutrophils and other white blood cells normally whiz through blood vessels at a high rate of speed (relatively speaking). In the vessels of an infected tissue, however, the neutrophils start to stumble and roll as the receptors on their surface latch on to the adhesive molecules expressed by the endothelial cells, a bit like runners in a road race suddenly encountering a patch of glutinous mud. Under the influence of the chemokines and other molecules that are released in response to the presence of microorganisms, the rolling neutrophils finally come to a dead stop as they attach firmly to the endothelial cells. They squeeze between these cells into the tissue, and then swarm to the site of infection as they follow a gradient of chemokines and other molecules that attract and guide them.

Once the neutrophils have arrived at the focus of infection, they fire their battery of antimicrobial weapons, many of which are similar to those used by macrophages, and then, having fulfilled their function, die shortly thereafter. Some of the neutrophils may engage in a last-ditch act of heroic self-sacrifice in which they spew out their nuclear and cytoplasmic contents—a mix of deoxyribonucleic acid (DNA) strands, chromosomal proteins, and antimicrobial peptides—to form a sticky net in which bacteria are trapped and eliminated. Unfortunately, several of the toxic molecules that are released by neutrophils against their microbial targets are also harmful to mammalian cells, so some degree of collateral tissue damage is unavoidable.

Monocytes and lymphocytes, the other major types of circulating white blood cells, are also recruited into infected tissue by the same broad mechanisms involved in neutrophils: rolling, firm adhesion, and migration out of the blood vessel under the guidance of chemokines. During the first few hours of the innate response, the chemokines produced in the tissue preferentially attract neutrophils, so those cells are usually the first to arrive from the circulation, especially during bacterial infections. Subsequently, the chemokine profile shifts, leading to the recruitment of monocytes, which differentiate into macrophages after they have left the blood and entered the tissue. In situations where adaptive immunity is activated, the response develops over the course of about a week, after which activated lymphocytes are also recruited and join the milling crowd of cells at the site of infection.

Inflammation: Stoking the Fire

The events that we have just discussed constitute an important aspect of innate immunity known as the acute inflammatory response,

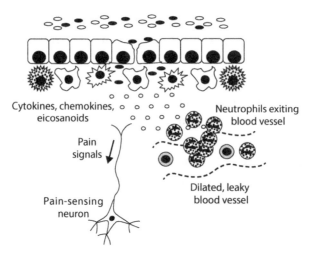

FIGURE 2.2. The inflammatory response, illustrated by the events that occur when an epithelium is breached by a significant number of microorganisms.

or inflammation (figure 2.2). Inflammation is something that all of us have experienced at one time or another, most commonly when a minor skin injury becomes infected. Within an hour or two, the affected area becomes red, warm, and swollen—reflecting increased blood flow and the movement of fluid, plasma proteins, and cells into the tissue— and also usually painful. This last symptom is the result of stimulation of nerve fibers, either by microbial products or by the cytokines that are released by macrophages, neutrophils, and other cells. The tetrad of redness, warmth, swelling, and pain—described by the first-century Roman scholar Celsus as *rubor, calor, tumor,* and *dolor*— represents the classic clinical manifestation of the local inflammatory response. Another occasional feature of inflammation is the formation of pus, the thick, yellowish fluid that sometimes accumulates in or oozes from infected tissues and that is made up of dead neutrophils, microbes, and cellular detritus.

In addition to the abnormalities that occur locally in the affected tissue, inflammation can have more widespread, systemic effects if it is severe enough. All of us have undoubtedly dealt with the unpleasant symptoms that often accompany a bad cold or other infectious illness and that we associate with the feeling of being sick: fever, distaste for food, and the desire to get away from it all and crawl into bed. The various sickness behaviors, as well as the fever, are manifestations of the underlying inflammatory process and are the result of the cytokines that are released at the site of infection. The cytokines make their way into the circulation and induce cells in the brain—endothelial cells and macrophage-like cells known as microglia—to secrete additional molecules, including small lipids, that act on specific nerve cells. These neurons are wired into neural circuits that control body temperature, appetite, mood, and social interactions. Depending on the particular network that they are connected to, activation of the nerve cells can make you unwilling to eat, reluctant to interact with other people, or unable to look on the bright side of life. Some of the activated circuits can also fool your brain into thinking that your body temperature is too low and that it needs to be raised by shivering and reducing blood flow to the skin (which is why you develop a fever).

The importance of fever and sickness behaviors in dealing with infection is not entirely clear, but there is some evidence to suggest that elevated body temperature may inhibit the growth of pathogens and increase immune cell activity. Social withdrawal may help to decrease transmission of pathogens from one individual to another, and loss of appetite can alter metabolism in ways that enhance the ability to tolerate the presence of certain types of pathogens. Some of these ideas are speculative or based on a limited number of animal experiments, so they are not fully substantiated. But what is undeniable is that fever and sickness behaviors have been conserved through evolution (they occur in most animals, and even cold-blooded ones like lizards and snakes, which cannot regulate their own body temperature, will move to a warm location when they are sick), suggesting that these responses probably have some survival value.

Resolution: Cooling Off

Once the microorganisms that initiated the inflammatory process have been eliminated, something that is usually accomplished only after the adaptive immune system has provided assistance, resolution starts to occur. The cell and microbial corpses and tissue debris that result from the battle between host and pathogen are ingested and digested by a cleanup crew of macrophages, which also facilitate healing by secreting molecules that promote tissue repair. The termination of inflammation results in part from the clearance of the microbe- and host-derived signals that kicked off the response, but in addition, there are several factors that actively suppress inflammation and induce the repair of damaged tissue. They include cytokines like IL-10, which are secreted by cells in the local environment; a subclass of eicosanoids known as specialized pro-resolving mediators (SPMs) produced by neutrophils, macrophages, and other cells; and probably additional molecules that are yet to be discovered. SPMs in particular help to reverse the effects of inflammation by inhibiting neutrophil recruitment while also promoting tissue regeneration and the removal of dead cells and debris. The action of SPMs and other anti-inflammatory and pro-resolution factors illustrates yet another theme in immunology: whenever a function of the immune system is turned on, mechanisms to turn it off when it is no longer needed are also deployed. These regulatory mechanisms ensure that the activity does not become excessive and tissue damaging; their failure can contribute to the development of disease.

Inflammation is a fairly stereotypical process that can occur in any tissue, and it is often identified by clinicians by simply adding the suffix "itis" to the name of the tissue. Thus, dermatitis indicates inflammation of the skin; carditis, inflammation of the heart; gastroenteritis, inflammation of the intestinal tract; and so on. With some variations, the underlying cellular and molecular events in each case are very similar. However, the health implications can be quite different depending on the exact cause of the inflammation, the tissue involved, and how long the process lasts. The short-lived, superficial skin inflammation that follows an abrasion is usually nothing more than a temporary annoyance, whereas carditis, whether caused by viruses, bacteria, or other agents, can have serious, sometimes life-threatening consequences because of the associated damage to the delicate structures of the heart. Moreover, inflammation in and of itself can be damaging to any tissue if it becomes chronic, either because the inciting agent cannot be cleared or because the mechanisms that normally lead to resolution are malfunctioning.

The innate immune response represents only the initial phase of antimicrobial defense, but it plays an essential role in keeping infection under control while additional protective mechanisms are being mobilized. Now that we have some idea about the cells, molecules, and processes involved in innate immunity, we will move on and try to answer an important underlying question: how exactly does the response get started?

CHAPTER 3

Innate Immunity II

Sensing Infection

The ability to distinguish "nonself"—what is alien to our bodies—from "self"—the constituents of our own tissues—is a fundamental attribute of the immune system, a characteristic that allows immune cells to mount an attack against external threats but prevents them from harming internal structures. In essence, immune activation occurs only in response to exogenous, potentially dangerous molecules and not to endogenous ones. Self-nonself discrimination is also a conceptual cornerstone of immunology, a common thread that is interwoven through much of the subject. It involves multiple mechanisms, some that operate during development, others during the actual course of the immune response. We will encounter several of them as we go along, and we will start by examining how cells of the innate immune system make the important distinction between microorganisms and mammalian cells.

Detecting the Enemy: Part I

Implicit in everything that we discussed in the last chapter is the idea that the presence of microorganisms must be sensed in some way before the appropriate responses are activated. Clearly, the sensing must involve the identification of something that is relatively unique to microbes and is not found on mammalian cells. In other words, it must be a physical feature or behavior that distinguishes the two and that acts as a signal indicating the presence of an intruder, a microbial intruder. The detection of this signal provides the green light for deployment of antimicrobial defenses.

In the case of innate immunity, an important mechanism of self-nonself discrimination that kicks off the response involves the recognition of molecules that are expressed mainly, if not exclusively, by microorganisms (figure 3.1). Such molecules are usually known as pathogen-associated molecular patterns or PAMPs, but they are also sometimes referred to by the more general term microbe-associated molecular patterns or MAMPs since they can be found on organisms that are not necessarily pathogenic. To avoid confusion, PAMP will be used consistently throughout this book even though the relevant molecule may originate from a nonpathogenic organism. Regardless of whether you call them PAMPs or MAMPs, the receptors that recognize these molecules are known as pattern recognition receptors or PRRs. (I have to acknowledge here that dealing with the alphabet soup of acronyms in which immunology is immersed can get tiresome at times, but the shorthand *does* help to convey information efficiently when discussing a complex subject. A list of the relevant abbreviations is provided at the end of the book for easy reference.)

One of the best characterized PAMPs is lipopolysaccharide (LPS), a combined lipid-sugar molecule that is an abundant component of the exterior coat of a major subset of bacteria. These LPS-expressing bacteria are classified as being Gram-negative since they fail to be stained by

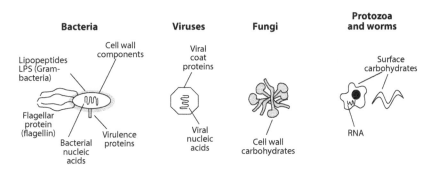

FIGURE 3.1. Pathogen-associated molecular patterns (PAMPs) of the main categories of microorganisms.

a method developed by the Danish bacteriologist Hans Christian Gram. LPS has long been known for its noxious effects on mammals, notoriety that has given rise to its alternate name endotoxin. Subsequently, it was also shown to induce macrophages, neutrophils, and other cells of the innate immune system to secrete the various cytokines, chemokines, and lipid mediators that contribute to inflammation. However, the molecular mechanisms involved in sensing LPS remained mysterious until the 1990s, when two separate lines of investigation converged and led to the identification of the long-sought LPS receptor.

Microbes Pay the Toll

The opening act of this drama, which started in the mid-1980s, had nothing to do with immunity. Rather, it was centered around a particular aspect of the embryonic development of the fruit fly, one of the workhorses of research in genetics. Investigators in Germany, led by Christiane Nüsslein-Volhard, were trying to understand how the top and bottom of the embryo were programmed to go through very distinct developmental changes, and they finally identified a specific gene that played a major role in this process. They christened this gene Toll. (There is an amusing, possibly apocryphal story associated with the name. It is said that when Nüsslein-Volhard first observed the bizarre embryo morphologies induced by mutations in this particular gene, she looked up from the microscope and exclaimed, *"Das war toll!,"* the German equivalent of "That's cool!" The description seemed apt and was duly enshrined in the gene's identity.)

About a decade later, Jules Hoffmann's group in Strasbourg, France, showed that Toll continued to function in the adult fly but that its role was very different from that in the embryo. Their studies demonstrated that adult flies with an inactivating mutation in the Toll gene rapidly succumbed to fungal infection. In a striking electron micrograph that accompanied the paper describing the work, the fungus was shown to have essentially taken over the bodies of the mutant flies, growing so rampantly that its hyphae protruded from the corpses like the tentacles of some alien life-form. These findings, which vividly illustrated the importance of Toll in antimicrobial defense, were soon followed by the identification of a whole family of Toll-like receptors (TLRs) in mammals, about a dozen in both mice and humans, and then the discovery by the laboratory of Bruce Beutler at the University of Texas Southwestern Medical Center in Dallas that one member of this family, TLR4, was responsible for recognizing and responding to LPS. Follow-up studies, carried out over several years by a number of investigators, clarified the structure of TLR4 and showed that it was embedded in the plasma membrane, with a large section or domain extending outwards into the environment outside the cell. Binding of LPS to this extracellular domain activated signals within the cell that led ultimately to the increased expression of multiple genes involved in antimicrobial defense and the inflammatory response. Other members of the TLR family had structures that resembled TLR4 and induced similar patterns of gene expression.

The landmark experiments on Toll and TLR4 represented the early stages of a renaissance in the field of innate immunity, a period of renewed interest and activity that extended the role of the initial phase of the immune response beyond the traditional one as first line of defense, and propelled it to a more central and vital position where, as we will see in the next chapter, it acted as an essential link in the activation of adaptive immunity. Hoffmann's and Beutler's contributions to this revival were recognized by the 2011 Nobel Prize in Physiology or Medicine (together with Ralph Steinman, whose work on further characterization of the connections between the innate and adaptive immune systems will figure later in our discussion). Nüsslein-Volhard and her colleague Eric Wieschaus had already received the Nobel Prize in 1995 for their work on fruit fly development.

Other PAMPs, Other Receptors

So, now we have some insight into how Gram-negative bacteria are sensed. But what about the bacteria that do not express LPS, the Gram-positive bacteria? Or fungi, viruses, and parasites? How are *they* all recognized by the innate immune system? Perhaps not surprisingly, further research has revealed that additional members of the TLR family help to detect a variety of other molecules that are found only on microorganisms. TLR5, for instance, recognizes flagellin, a protein subunit of the flagellum, a whip-like appendage that propels the movement of certain types of bacteria, while TLRs 1, 2, and 6 in different combinations respond to the presence of lipopeptides, greasy molecules found in the cell walls of both Gram-negative and Gram-positive bacteria. TLRs 3, 7, and 8 are triggered by viral or bacterial ribonucleic acids (RNAs), while TLR9 recognizes bacterial DNA. Some of the TLRs that are expressed in mice have been lost during the course of evolution and are not found in humans. TLR11 and TLR13, which recognize a protozoan protein and bacterial RNA, respectively, are examples of such species-specific members of the family. Presumably, their functions are not as important in humans as in mice or are carried out by other TLRs.

In most of these examples, as in the case of the TLR4-LPS interaction, the binding of the PAMP to the extracellular domain of the TLR activates signals that ultimately bring about specific behaviors in the responding cell, including the deployment of antimicrobial mechanisms and the expression of cytokines, chemokines, and other molecules involved in inflammation. Thus, the TLRs by themselves can sense many of the microbes that cells of the innate immune system are likely to encounter and can induce the appropriate protective responses. However, since their sensing domains are directed toward the outside of the cell, they can only detect microbial molecules present in the extracellular environment or in material taken up from that environment into the interior of a phagosome or an endosome (the latter being a smaller version of the former). This limitation is significant since viral pathogens spend most of their infectious cycles replicating in the cytoplasm of the cell, and even some bacterial pathogens end up in that location. Defending against such organisms requires a microbial-sensing mechanism within the cytoplasmic compartment.

Just as the TLRs represent key sensors of microbes in the extracellular environment, there are PRRs that play a similar role in the cytoplasm. An important group of these cytoplasmic PRRs forms what is referred to as the inflammasome family. Each member of this family is made up of multiple proteins that assemble into a complex structure known as the inflammasome when activated by the appropriate stimulus. Different types of inflammasomes respond to different microbial molecules, including LPS, flagellin, and certain virulence proteins expressed by some bacterial pathogens, as well as to nonmicrobial stress stimuli associated with changes in cytoplasmic ion concentrations or the presence of reactive oxygen radicals. Once the inflammasome is activated, it functions as an enzyme that cleaves a precursor of the cytokine IL-1β as well as other proteins that facilitate secretion of the mature cytokine. The release of IL-1β acts as a signal, a red flag, that indicates that the cell is infected or otherwise stressed, and it initiates or amplifies an inflammatory response that helps to deal with the situation.

What about viruses in the cytoplasm? One way of identifying the presence of these pathogens is through the nucleic acids, DNA or RNA, that either represent their genomes or that are generated during their growth within the cell. There are a number of different PRRs in the cytoplasm that detect viral DNA or RNA and then trigger signaling cascades leading to the activation of defensive mechanisms. Many of these signals promote the expression and secretion of type 1 interferons, which act on surrounding cells to increase their resistance to viral infection. Some of the signals also induce the expression of cytokines such as TNFα and IL-6, which contribute to the inflammatory response.

Epithelial cells, macrophages, neutrophils, and other cell types involved in innate defense typically express several kinds of PRRs. The receptors are deployed strategically at the cellular locations most appropriate for surveying the environments—extracellular, phagosomal and endosomal, and cytoplasmic—where pathogens are likely to be encountered (figure 3.2). The PRRs often include representatives of the TLR, inflammasome, and cytoplasmic nucleic acid sensing families. In addition, other types of PRRs also contribute to innate immunity by detecting microbial molecules that both overlap with and extend beyond those recognized by these three main families. For example, there is a large family of PRRs that sense and respond to unusual

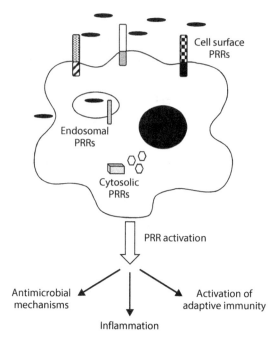

FIGURE 3.2. The categories and distribution of pattern recognition receptors (PRRs) expressed by a typical innate immune cell and the effects of their activation.

carbohydrate structures present on the surface of fungi, protozoa, and parasitic worms. The expression of a repertoire of different PRRs allows the cell to "see" most microbial intruders regardless of whether they are outside or within, and also confers a degree of redundancy that provides backup functions if any one detection system fails.

Broadly speaking, and summarizing some of what has been mentioned already, the signals and gene expression changes that are induced by most PRRs lead to three main types of functional outcome (figure 3.2). Two of them—the triggering of cellular antimicrobial mechanisms and the activation of inflammation—have been discussed in the last chapter. The third involves communicating with and activating the adaptive immune system and will be the subject of a later chapter. Individual PRRs may elicit responses that interact with each other and lead to variations on these three themes. As a result, the final outcome may depend on the combination of PRRs that are triggered, which, in turn, can vary with the microorganism that is involved. Microbe-specific differences in the pattern of PRRs that are triggered may allow a limited diversification of the kind of innate immune response that develops, which could be helpful in dealing with different types of pathogens. Finally, it should be mentioned that some surface-located PRRs act largely, if not exclusively, to facilitate phagocytosis of microorganisms.

Detecting the Enemy: Part II

The mechanisms of microbial sensing discussed so far depend on detection of the organism itself based on the PAMPs that they express. But microbes are clever little beasts, and in some cases they have evolved ways to evade immune defenses by masking or disguising their PAMPs. This would render them "invisible" were it not for the fact that the immune system has another strategy for identifying the incursion of microorganisms, and that is by detecting the cell damage that they cause. Injured cells become leaky and, consequently, they start to release molecules that are normally sequestered within intracellular structures into the surrounding environment. When such molecules—nuclear or cytoskeletal proteins, for instance, or adenosine triphosphate (ATP), the energy currency of the cell—are found outside cells, they constitute an indicator of cell damage and are collectively referred to as damage-associated molecular patterns, or DAMPs. DAMPs function very much like PAMPs: they are detected by some of the same PRRs (although there are also PRRs that are specialized for DAMP recognition), and they activate the same kinds of responses. They are thus an alternate means of alerting the immune system to the presence of microorganisms, particularly those that are likely to be harmful, and they also help to amplify the inflammatory events initiated by PAMPs.

Cells and tissues can be damaged even in the absence of microbial invaders, of course—because of trauma, for example, or during surgery. The release of DAMPs in such circumstances can lead to a sterile inflammatory process that has many of the clinical hallmarks of the inflammation that occurs during infection, something that is quite obvious from the appearance of a badly sprained ankle. The underlying molecular and cellular characteristics are also quite similar. Apart from trauma and surgery, sterile inflammation is involved in chronic autoimmune diseases such as rheumatoid arthritis and lupus, and may be triggered or perpetuated by the release of DAMPs.

Friend or Foe?

It is worth emphasizing here that PRRs do not distinguish between members of the resident microbiota and pathogens; they simply indicate the presence of a microorganism. TLR4, for instance, will respond as well to the LPS of a harmless strain of *Escherichia coli* that is a part of the gut microbiota as it will to the LPS of *Salmonella enterica* serovar Typhimurium, a pathogen that causes acute gastroenteritis. The distinction between nonpathogen and pathogen is based to a great extent on the virulence properties and behavior of the organism, not just on the PAMPs that it displays. The microbiota strain of *E. coli* does not usually provoke an inflammatory response since it resides outside of the intestinal epithelial barrier and so does not come in contact with TLR4-expressing cells in significant numbers (generally speaking, the epithelial cells themselves do not express PRRs on the side that faces the microbe-rich lumen of the gut). In contrast, *S.* Typhimurium has evolved highly specialized virulence mechanisms that allow it to invade actively through the epithelium and grow robustly within macrophages on the other side. Once the *Salmonella* bacteria have crossed the epithelium, their LPS molecules are immediately sensed by macrophages and other cells that express TLR4, leading to the activation of a vigorous intestinal inflammatory response that manifests as abdominal pain, nausea, vomiting, and diarrhea. A similar response could occur if a sufficient number of the *E. coli* happened to get through a weakened epithelial barrier.

Segregation by the epithelium is not the only way by which the microbiota is prevented from activating innate immunity. Some bacterial members of the microbiota have developed structural modifications of their LPS or other PAMPs that weaken or completely prevent recognition by PRRs. Most nonpathogenic organisms also do not cause cell damage, so they do not trigger the release of DAMPs that might amplify the inflammatory response, whereas infection by pathogens is typically associated with DAMP release. Moreover, some of the virulence molecules that allow pathogens to invade host cells and resist antimicrobial mechanisms trigger inflammatory responses because they are sensed by certain PRRs, including members of the inflammasome family. Finally, tissues with epithelial surfaces that are exposed to a dense microbiota community—the colon, for example—often have high concentrations of immune cells and molecules that dampen inflammation. The threshold for activating the inflammatory response is correspondingly elevated in such tissues, making it less likely that nonpathogenic organisms would provoke an inflammatory process severe enough to cause damage even if a few somehow made their way across the epithelial barrier.

Complementary Aspects of the Innate Immune Response

Our discussion would not be complete without mentioning two additional components of innate immunity. The first is a group of cells known as innate lymphocytes, which includes natural killer (NK) cells and innate lymphoid cells (ILCs). As their collective name suggests, these cells have some of the characteristics of lymphocytes. Unlike B and T lymphocytes, however, NK cells and ILCs do not have receptors for antigen. Instead, they are activated by the cytokines and other molecules that are expressed in the context of infection and cell damage, and they participate in the innate immune response by rapidly deploying specific effector mechanisms. NK cells are found in multiple tissues, including in the blood, and they kill virally infected or malignant cells, either directly or with the assistance of antibodies. They can also secrete cytokines, mainly interferon γ. ILCs are particularly abundant in mucosal tissues and contribute to the early stages of immune defense by secreting a variety of cytokines and other molecules. They are classified into three main subsets based on patterns of surface protein and cytokine expression, with each subset playing distinct roles in the innate response.

The other function that is necessary to mention here relates to the humoral or noncellular component of innate immunity, one aspect of which involves several soluble PAMP- and DAMP-recognizing proteins that are secreted by various cells, including liver cells, during a systemic inflammatory response. These proteins make their way into the blood and tissue fluids and bind to their targets, usually via specific carbohydrate molecules. Once bound, they activate another important module of humoral innate immunity known as complement. Complement is the collective name for a group of plasma proteins that circulate in the blood in an inactive form. These proteins can be activated in a serial cascade as part of innate defense, and also by some types of antibodies during the adaptive immune response. The activation of complement helps to clear infection in several ways: certain activated complement proteins coat (or opsonize) microbes and facilitate their phagocytosis by macrophages and neutrophils, while others amplify the inflammatory response or directly kill specific kinds of bacteria.

The first battle of a long campaign is now under way. The intruders have been detected and engaged but probably not completely defeated. In all but the most minor or transient of infections, the activation of innate immunity also initiates a second phase of the immune response, adaptive immunity. It is the combined and coordinated action of these two major arms of the immune system that usually results in the successful eradication of invading microorganisms and a return to health.

CHAPTER 4

Adaptive Immunity

Strength in Diversity

The innate immune response serves the important function of containing microbial invaders during the first minutes, hours, and days of infection. However, it has weaknesses that compromise its ability to clear infections completely except in the most minor cases. The PRR repertoire expressed by cells of the innate immune system is restricted in its diversity and allows the recognition of only broad categories of PAMPs and DAMPs. Moreover, the responses activated by most PRRs are stereotypical in nature and do not vary a great deal with the type of microorganism encountered. In general, they also do not change on repeated encounter with the organism. Many microbial pathogens have exploited these and other shortcomings of the innate immune response and have evolved mechanisms to evade or overcome its somewhat restricted defensive armamentarium. Because of the inherent limitations of the innate immune system, the eradication of most pathogens requires the assistance of the second phase of the immune response, adaptive immunity.

Variety in Form and Function

The chief orchestrators of the adaptive immune response are B lymphocytes and T lymphocytes, named for their development, respectively, in the mammalian bone marrow and the thymus, the latter being a rather nondescript organ located behind the breastbone. (Note, however, that the earliest progenitors of T cells originate in the bone marrow and then migrate to the thymus to complete later stages of maturation; additionally, if one is being a stickler for historical accuracy, it is worth mentioning that B lymphocytes actually derive their name from their development in an organ called the bursa of Fabricius, which is found in chickens and some other types of birds.) B cells and T cells have two special characteristics that make them particularly effective at fighting infection. The first is the ability to perceive an almost infinite variety of molecules (known as antigens when they are the targets of recognition by B or T lymphocytes), which enables them to mount a response to just about any microorganism that might be encountered. This ability is conferred by specific antigen receptors, the B cell receptor (BCR) expressed on the surface of B lymphocytes, and the distinct, although structurally related, T cell receptor (TCR) expressed on the surface of T lymphocytes (figure 4.1). BCRs and TCRs have highly variable antigen-recognition regions, each of which can bind to a unique molecular structure or to very closely related structures. Importantly, all the antigen receptors expressed by an individual B or T lymphocyte are identical in structure and specificity of recognition, but they differ from one lymphocyte to the next, features that make these receptors "clonotypic" in nature (contrasting with the "non-clonotypic" PRRs involved in innate immunity, which do not

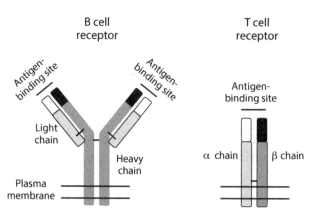

FIGURE 4.1. The B and T cell antigen receptors. The constant regions are shown in shades of gray, and the variable regions in white and black. The short black lines connecting polypeptides indicate disulfide bonds.

differ between individual cells). The total number of different antigen receptors expressed by all the B and T cells in a single human or other mammal runs into the millions for each of the two lymphocyte populations, providing the enormous recognition power needed to deal with the universe of foreign antigens, including those associated with microbes, that the immune system might be exposed to.

The second characteristic of B and T lymphocytes that makes them so good at warding off microbial attack is the diversity and versatility of their responses. The two types of lymphocytes have distinct but complementary functions that dispose of foreign antigens in different ways: B cells represent the humoral arm of adaptive immunity, which is mediated by secreted forms of BCRs (antibodies, also known as immunoglobulins, abbreviated as Ig) that make their way into the circulation and tissue fluids, while T cells are responsible for the cell-mediated arm, which involves eliminating antigenic targets either directly or by coordinating the action of multiple cell types. Within these two broad sets of mechanisms, additional specialization of function is possible based on the generation of different antibody and effector T cell types. Specific combinations of the various B and T cell functions can be activated depending on the nature and the context of the antigenic challenge, allowing the deployment of distinct types of adaptive responses against, for example, viral and fungal pathogens, or even against individual pathogens of the same class, such as the bacteria *Salmonella* and *Staphylococcus*. The exquisite specificity of antigen recognition and the ability to exert a variety of functional effects endow the adaptive immune system with a finely tuned sophistication that is lacking in the innate immune response. However, as we will see in more detail later, both humoral and T cell-mediated responses ultimately have to work together with elements of innate immunity to eliminate the foreign antigen or microorganism.

The multiplicity of antigen receptors and functional capabilities makes the adaptive immune response a potent weapon against active, ongoing infection. In addition, there is a third characteristic of B and T cells—memory, the ability to remember past antigen encounters—

that gives them the ability to defend against *future* infection. Immunologic memory, as it is traditionally described, is the consequence of an initial exposure to an antigen that manifests as a significantly faster and more robust B and T lymphocyte response when subsequently encountering the same antigen but not a different one. Or, to emphasize the key point, it is the heightened responsiveness of B and T cells when the same antigen is seen again. Memory is the result of the emergence during the initial adaptive response of a population of antigen-specific B and T lymphocytes that have been reprogrammed to survive for many years and to respond more quickly and with greater magnitude to the antigen that originally activated them. These memory lymphocytes continue to circulate through the blood and tissues long after the original antigenic challenge has been cleared, and they are available for rapid and robust activation if the same antigen is ever seen again. The ability to generate a population of memory lymphocytes makes excellent biologic and evolutionary sense: if a microbial pathogen has caused an infection once, there is a good chance that it will do so another time; being able to respond to it more effectively during a subsequent encounter could prevent or minimize the damage that it might cause. As we shall see in a later chapter, it is the power of immunologic memory that has been harnessed to great effect in vaccination strategies.

Diversity of antigen recognition, diversity of function, and memory: these are three fundamental attributes that are at the heart of adaptive immunity. We will learn more about each in the discussions that follow.

Recognizing the Other

Initial clues about the way lymphocytes recognize antigens came from studies on the structure of antibodies, which as we learned earlier, are essentially soluble, secreted forms of the BCR. These pioneering investigations were carried out independently by Gerald Edelman of the Rockefeller University in New York and Rodney Porter of Oxford University in the United Kingdom. Edelman made use of myelomas, which are antibody-secreting tumors that develop from the B lymphocyte lineage. Importantly, each myeloma originates from a single, malignantly transformed cell and so produces a population of identical antibody molecules that are directed against a single, usually undefined, antigen. By analyzing the structure of purified, myeloma-derived antibodies, Edelman discovered that each antibody molecule consisted of two identical large proteins (called immunoglobulin heavy chains) and two identical small proteins (called immunoglobulin light chains). The proteins were linked by chemical bonds to form a complex in which the heavy and light chains were paired with each other. (Figure 4.1, *left*, shows the structure of the BCR, which is the membrane-bound form of an antibody.) He also found that each heavy- and light-chain protein contained a region that was relatively constant in structure, as well as a variable region that differed quite markedly depending on the myeloma from which the antibody was obtained. Porter came to similar conclusions based on his studies of rabbit antibodies. The findings of these parallel lines of investigation suggested that it was the variable regions of the heavy and light chains that together formed the antigen-binding part of the antibody, and, by extension, that antibodies with different variable regions allowed the recognition of different antigens. Since each antibody molecule was made of two identical heavy-light pairs, two antigen-binding sites were formed by the combination of the variable regions of the two types of proteins. The insights into antibody structure provided by the Edelman and Porter experiments were subsequently confirmed, extended, and refined by detailed crystallographic analyses.

The TCR was identified by DNA cloning techniques only in the 1980s, long after Edelman's and Porter's work on antibodies. But there was a satisfying similarity between the structures of the TCR and the BCR (see figure 4.1). Although the TCR was made up of only two proteins, one α chain and one β chain (or one γ chain and one δ chain in the case of a minor subset of T cells), each had variable and constant regions that resembled the corresponding parts of the heavy- and light-chain proteins of the BCR. The parallels suggested that the two types of receptors recognized antigen in much the same way, with the variable regions of the component proteins contributing to the formation of the antigen-binding site.

The Diversity Conundrum

While the structural characterization of antibodies was a major milestone in immunology (one that earned Edelman and Porter the 1972 Nobel Prize in Physiology or Medicine), it also represented something of a puzzle. Experiments performed by a number of immunologists during the first half of the twentieth century had shown that antibodies could be raised against a multitude of different antigens, even ones that did not exist in nature. If individual antigens were recognized by antibodies with distinct variable regions, how could the vast number of different antibody heavy- and light-chain proteins required to explain the almost infinite capacity for antigen recognition be encoded in the genome? If one went by accepted dogma and assumed that each of those proteins was derived from a single unique gene, then the genome would be taken over and overwhelmed by all the antibody genes and there would be no room to encode everything else needed for life.

This was the conundrum that Susumu Tonegawa was trying to resolve in the 1970s when he was at the Basel Institute of Immunology in Switzerland. His approach was based on a painstaking analysis of the bits of genomic DNA that corresponded to the parts of the antibody molecule identified by Edelman and Porter. Over the course of several years, he and his colleagues made a series of discoveries that ultimately pointed to an unprecedented mechanism for generating antibody diversity.

To understand this mechanism, it's useful to start by playing a little game (in your mind, at least, if not in actual fact). Take a traditional deck of cards, with its four suits: clubs, spades, diamonds, and hearts. Place all the red suit cards in a pile at the left end of one side of a table, and the clubs and spades in the middle and the right, respectively. Choose any one of the 26 different red cards (which we will denote as R) followed by one of the 13 different clubs (C) and then one of the 13 different spades (S), and line them up right next to each other in the center of the table to form the sequence R-C-S. How many different R-C-S sequences can you make with the cards you have available? A bit of simple math will tell you that the answer is $26 \times 13 \times 13 = 4,394$. So, from a relatively small number of cards (52), you've been able to make almost 4,500 different R-C-S sequences—very impressive! Developing B cells use a similar strategy to generate an even greater diversity of BCRs.

From Card Shuffling to Genetic Shuffling

The work carried out by Tonegawa and his collaborators led ultimately to the surprising conclusion that the DNA encoding the variable regions of the heavy- and light-chain proteins of the BCR consisted of multiple segments: V, D, and J for the variable regions of the heavy-chain gene (located on chromosome 14), V and J for the variable regions of the light-chain genes (of which there are two—κ and λ—located on chromosomes 2 and 22, respectively). The gene segments started out being widely separated on their respective chromosomes but, astonishingly, they were rearranged and brought next to each other in the final configurations V-D-J and V-J for the heavy- and light-chain genes, respectively (figure 4.2).

The process that Tonegawa discovered is essentially a genomic version of our card game, a shuffling of DNA that occurs in developing B lymphocytes but not in other cell types (with one exception, as we will see below). Like the different cards of each category, there are multiple versions of each gene segment that can be used to encode the finished variable region—thirty to forty V gene segments, twenty or so Ds, and five or six Js—except that they are spread out in a linear array rather than being piled on top of each other like the cards. The version of each segment that is chosen for putting together the variable region-encoding DNA is more or less random in individual B cell progenitors, with the result that each mature B cell ends up with unique

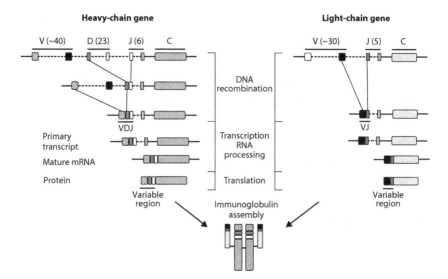

FIGURE 4.2. The steps involved in the generation of the B cell receptor: Recombination of the DNA encoding the immunoglobulin heavy- and light-chain variable region gene segments, followed by transcription and RNA processing, translation of the mature RNA to produce the heavy- and light-chain proteins, and assembly of the two protein types to form the complete complex.

DNA rearrangements with the ability to give rise to distinct heavy- and light-chain variable regions. Based on the number of different Vs, Ds, and Js available, the diversity of heavy- and light-chain variable regions generated in the population of mature B cells by random recombination of these segments is on the order of 5,000 (40 Vs × 20 Ds × 6 Js) for the former and 150 (30 Vs × 5 Js) for the latter. For both the heavy and light chains, the cluster of J gene segments is linked by an intervening sequence or intron to one or a few DNA segments encoding the constant region (C). Following the recombination events that result in the generation of the DNA encoding the variable regions, regulatory elements adjacent to the recombined V segment drive the production of a primary or immature RNA transcript encoding the variable and constant regions. The noncoding portion (intron) of this transcript is spliced out to form the mature messenger RNA (mRNA), which is then translated into the heavy- or light-chain protein (figure 4.2).

The recombination of V, D, and J gene segments is not the only mechanism of diversification of the heavy- and light-chain variable regions. An even more important source of diversity is generated at the junctions between these segments by the random addition or deletion of nucleotides in individual developing B cells, a process that amplifies the number of different variable regions that can be produced manyfold. Moreover, since any heavy chain can pair with any light chain, the potential diversity of the BCR and antibody repertoire generated by the random combinatorial, junctional, and pairing mechanisms is theoretically in the billions, more than enough to recognize just about any antigen that might ever be encountered.

The recombination of DNA that gives rise to the BCR is highly regulated. It starts at the heavy-chain genes (one gene in each parental chromosome 14) with a D to J rearrangement, followed by a V to D-J. Only after the process has succeeded in producing a functional protein does the developing B cell go on to rearrange the light-chain genes. This ordered sequence of events helps to conserve valuable biological resources: a relatively high rate of failure is intrinsic to the recombination process and there would be no use in a cell rearranging light-chain genes if it does not have a heavy-chain protein. B lymphocyte progenitors that fail to make a heavy or light chain are allowed to die, avoiding wasting space and nutrients on these nonfunctional cells.

There is an additional important regulatory mechanism that is involved in B cell development. As soon as rearrangement at one heavy-chain gene has succeeded in producing the corresponding protein, further rearrangements of the heavy-chain gene on the other chromosome 14 are immediately stopped. This cessation of recombination, known as allelic exclusion, also operates during rearrangement of the light-chain genes, ensuring that an individual B lymphocyte expresses BCRs that all consist of one unique type of heavy chain and one unique type of light chain, and so has the ability to recognize only one antigenic target.

DNA is generally thought to be a stable repository of genetic information, immutable except in the germ cells that give rise to sperm and ova. That it could be shuffled and moved around in a somatic cell type such as the B cell challenged conventional biological thinking at the time that Tonegawa made his observations. But the process of somatic recombination (also known as VDJ recombination in the present context) provided a neat and satisfying explanation for how a limited amount of DNA could be used efficiently to produce the enormous diversity of BCRs and antibody molecules required to recognize all the foreign antigens that the immune system might have to deal with. Elucidation of the process led to a Nobel Prize for Tonegawa in 1987 and, importantly, subsequent work showed that a virtually identical somatic recombination mechanism was also responsible for producing the diverse repertoire of antigen receptors in T lymphocytes.

Interlude in RAGtime

The story of VDJ recombination has an epilogue that is worth mentioning, in part because I have a personal connection to it but also because it nicely illustrates some of the vagaries of scientific research, especially the roles played by audacity, persistence, and serendipity. One of the big unresolved questions about the somatic recombination process was how exactly the rearrangement of gene segments occurred during the development of B and T cells. To answer this question, it was first necessary to identify the proteins that carried out the cut-and-paste functions needed to move specific pieces of DNA from one part of the genome to another so that they joined up to code for a BCR or TCR protein.

David Schatz, a young MD-PhD student, was tackling the problem in the laboratory of David Baltimore at the Whitehead Institute in Cambridge, Massachusetts, at the time that I was just beginning postdoctoral training in another lab there. Rather than taking the traditional and very laborious approach of biochemically purifying the proteins involved in VDJ recombination, Schatz decided to pursue an ambitious strategy to directly clone what he hoped would be the key gene involved in the process. He described the strategy at an in-house symposium soon after I started at the Whitehead, and I remember joining the crowd around his poster to listen to his explanation. His idea was to introduce fragments of genomic DNA into a fibroblast, a generic cell type found in most tissues that is normally incapable of executing somatic recombination, with the hope that the forced expression of a gene on one of the fragments would induce the cell to behave like a B lymphocyte precursor and rearrange a reporter DNA construct with some of the features of BCR gene segments. To make his task a little easier, Schatz designed the reporter so that its successful rearrangement would allow the fibroblast to survive and proliferate in the presence of a specific antibiotic. If the cells were cultured in medium containing the antibiotic after introducing the genomic DNA, only those cells that received a fragment with a gene capable of inducing VDJ recombination would grow, while all the others would be killed. By isolating and sequencing the introduced DNA from the surviving fibroblasts, it should be possible to identify the relevant gene.

The scheme was very, very clever . . . at least in theory. While Schatz was enthusiastically proclaiming the virtues of his approach, it was obvious even to a novice like me that the plan was fraught with risk. I remember that one postdoc in the crowd went so far as to snigger disdainfully, "This will never work," as he moved on to the next poster. The skepticism was probably justified: not only were the technical challenges formidable, but the success of the entire procedure depended on there being a single gene that could kick off the whole recombination process. Basically, it was the biological equivalent of the alchemist's dream of turning lead into gold. But Schatz was apparently undeterred by doubt, either his own or others', and he forged ahead, even in the face of gentle discouragement from his boss.

Fortune favors the brave, they say, and so it proved in this case. Against all odds, Schatz's strategy worked, and he succeeded in obtaining fibroblasts that had been induced to rearrange the reporter DNA. But it was an additional three years of twelve-hour days and multiple setbacks before he and another student, Marjorie Oettinger, isolated the DNA fragment with the characteristics they were looking for. When they sequenced it, they found not one gene but two separate ones, which they named recombination activating genes (RAGs) 1 and 2. Follow-up experiments showed that both of the genes were required to induce VDJ recombination, which meant that the stars must have been aligned in a particularly auspicious configuration when Schatz started the project, because if it had not been for the fact that the genes were relatively small and placed close together, his strategy would have never worked. That stroke of luck made his and Oettinger's careers. They both landed faculty positions immediately after finishing their PhDs (a very unusual achievement since most students are expected to go through a period of postdoctoral training first) and established productive, independent labs focused on the detailed mechanics of how exactly RAG1 and RAG2 functioned.

Searching for Foes without Harming Friends

Somatic recombination is a very effective mechanism for generating the diversity of antigen receptors needed by B cells and T cells to recognize the countless numbers of foreign molecules that the immune system might have to respond to. However, the quasi-random nature of the process means that among the many different antigen receptors that are expressed by developing B and T lymphocyte populations, there will be some that react with self molecules found on the individual's own cells. The existence of such autoreactive lymphocytes, which have the potential to mount damaging responses against any or all tissues, would seem to violate the core principle of self-nonself discrimination on which immune function is based. However, developing B and T lymphocytes are subjected to a censoring mechanism known as negative selection, which removes cells with the ability to recognize self antigens. During negative selection, any developing lymphocyte with an antigen receptor that binds specifically to an endogenous (self) molecule present in the local tissue environment is eliminated as a result of signals delivered through the receptor. The elimination of such self-reactive lymphocytes involves either programmed cell death (which can occur in both B and T cells) or the replacement of the initial receptor with one that does not recognize self molecules (a process known as editing, which occurs only in B lymphocytes and is the dominant censoring mechanism in these cells). Negative selection enforces self-nonself discrimination by ensuring that the mature B and T lymphocyte populations that emerge from the bone marrow and thymus, respectively, are largely depleted of any self-reactive cells and are directed almost exclusively against foreign antigens. The process is not perfect, however, and additional mechanisms are involved in preventing the development of autoimmunity.

Our lymphocyte army, with potential traitors identified and purged, is now ready to march out in search of the enemy. Approximately 10–20 million newly formed B cells and about one-tenth that number of newly formed T cells enter the bloodstream every day. These cells are usually described as being naïve since they have not yet encountered antigen. They embark on a blood-borne circumnavigation of the body that takes them sequentially through the peripheral lymphoid organs—the spleen and various lymph nodes—a migration pattern that is designed to maximize their chances of detecting any foreign antigens that might be present. At each of the lymphoid organs, the naïve lymphocytes exit the circulation and remain in the tissue for a few hours as they scan for the presence of an antigen that can interact with their BCRs or TCRs. If the appropriate antigen is not encountered, the cells return to the circulation (via a rather circuitous detour through lymphatic vessels in the case of exit from the lymph nodes) and continue their patrol until either the right antigen is detected or their life span is completed, which usually occurs in a few days. The lost cells are, of course, replaced by newly generated ones that emerge from the bone marrow and thymus, maintaining the numbers of B and T cells at a relatively constant level under normal conditions.

Strategic Lymphocyte Migration

You might think that it would make more sense for the naïve B and T cells to migrate through peripheral tissues such as the skin, gut, or airway, since those are the areas where foreign material, including microorganisms, are most likely to enter the body. However, there is a reason that they have evolved mechanisms to migrate through the spleen and lymph nodes instead. The spleen acts as a filter for all the blood flowing through the body, so any foreign antigens that make their way into the blood must necessarily pass through this organ. Lymph nodes act similarly at a regional level. For example, the tissue fluid (lymph) from the whole leg drains through lymphatic vessels into the lymph nodes of the groin and brings with it antigenic material from every part of the limb, either free-floating or carried by cells. Similarly, the lymph nodes in the armpit receive lymph from the whole of the arm. So, in terms of monitoring the body for the presence of foreign antigen, it is decidedly more efficient for the lymphocytes to check the spleen and lymph nodes rather than spending a lot of time and energy wandering through all the peripheral tissues. There is another advantage to the former strategy. Only a very small fraction of the total naïve B or T cell population carries an antigen receptor that is able to recognize a specific foreign antigen, about one lymphocyte out of every hundred thousand to a million. The architecture, cell composition, and relatively restricted space of the lymphoid organs increases the probability that the antigen will actually meet up with the right lymphocyte.

At this point, our discussion of B and T cells must diverge since the two populations are activated in different ways and carry out distinct functions. We will begin with B lymphocytes, remembering, of course, that both the humoral and cell-mediated arms of the adaptive immune response are generally activated and deployed in parallel.

CHAPTER 5

B Lymphocytes

Antibodies "R" Us

Consider the case of a young boy who has injured his left thumb. Unfortunately, the affected area has become contaminated with a virulent strain of the common skin pathogen *Staphylococcus aureus*, resulting in the activation of innate immunity and a vigorous inflammatory response. The thumb has become red, swollen, and painful, and it feels warm to the touch.

If we were endowed with a superpower that allowed us to take a look inside the infected tissue, we would see that the resident macrophages, aided by an influx of neutrophils from the circulation, have killed some of the bacteria, although many are multiplying rapidly. We could track fragments of the killed bacteria and even whole organisms as they make their way into the lymphatic vessels of the inflamed thumb and float along in the lymph draining from the tissue until they wash up in the lymph nodes in the left armpit. We may also see bacteria or their degradation products being carried through the lymph vessels by dendritic cells, which, as you may recall, are tissue-resident phagocytic cells that convey microbial and antigenic material from the periphery to the regional lymph nodes. When the mix of bacteria, bacterial detritus, and cells reach the lymph nodes, a local inflammatory response is induced as a result of PRRs on innate cells being activated by staphylococcal PAMPs. The nodes will probably become swollen and painful as a consequence. Importantly, the B cells in the node will be exposed to staphylococcal antigens, in soluble form, on the surface of whole bacteria, or brought by dendritic cells. Among the millions of naïve B cells in the node that are passing through as part of their routine patrolling function, there will be a handful that have BCRs with the ability to bind to the antigens. These cells will be chosen, or "clonally selected," for activation by the binding of the staphylococcal antigens to the corresponding BCRs.

B Cell Activation: With or without Help

Depending on the nature of the antigen, B cell activation can occur in two distinct ways. If the antigen has multiple, structurally identical components, each of which can bind to one of the hundreds of identical BCR molecules on the surface of an individual B cell, the interaction will cause multiple BCRs to aggregate and deliver a robust intracellular signal (figure 5.1, *left*). The strength of this signal is sufficient on its own to induce the cell to start multiplying. This mode of B cell

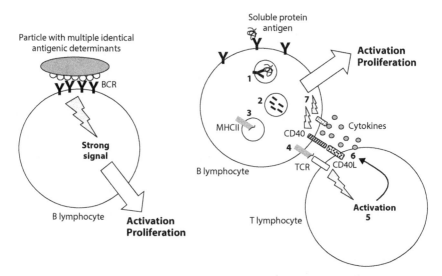

FIGURE 5.1. B cell activation mechanisms: T-independent (*left*) and T-dependent (*right*). In T-independent activation, the aggregation of B cell receptors (BCRs) by the binding of multiple identical antigens delivers a strong signal that is sufficient to drive B cell activation and proliferation. T-dependent activation, which involves a soluble antigen binding to a single BCR, requires multiple, sequential steps (indicated by the numbers in the figure) that ultimately lead to B cell activation and proliferation.

activation is known as T-independent activation since it does not require the assistance of T cells and is typically brought about by molecules with many repeating subunits—carbohydrate or lipid components of bacterial cell walls, for example.

The other mode of activation involves antigenic molecules that bind only one BCR each on an individual B cell (see figure 5.1, *right*). This type of interaction is characteristic of soluble protein antigens. It does not cause BCR aggregation and does not deliver a strong signal. Instead, the BCR with the bound antigen is internalized (endocytosed) by the B cell into an intracellular compartment, the antigen is degraded, and its peptide fragments are loaded onto a protein called the class II major histocompatibility complex (MHC) molecule and displayed on the cell surface. The MHC molecule is also known as the human leukocyte antigen (HLA) in members of our species. The antigen peptide-MHC duo can be recognized by a CD4-expressing helper T cell that has the appropriate TCR and that has already been activated by a dendritic cell presenting the same combination of peptide and MHC (we will learn more about this process and about helper T cells in the next chapter). When such a pre-activated T lymphocyte detects the displayed peptide on the surface of a B lymphocyte, it provides "help" in the form of various molecules that it is induced to express. The B cell senses these molecules, including a surface protein called CD40 ligand (CD40L) and secreted cytokines like IL-4 and IL-21, via corresponding receptors (CD40 and specific cytokine receptors) and is stimulated to undergo proliferation. For obvious reasons, this form of B cell activation is described as being T-dependent.

This is probably a good place for a brief digression on CD (cluster of differentiation) molecules, which are hard to avoid in discussions of immunology, including in the paragraph above. These molecules, which are also known as CD antigens, are present on the surface of cells and can be detected there using defined antibody reagents. Most of them (but not all) are proteins and their expression varies with cell type. For instance, CD3 is found only on T lymphocytes, while CD19 is characteristic of B cells. Their cell type-specific patterns of expression and relative ease of detection make CD antigens convenient markers of cell identity. They are widely used for this purpose during the characterization of specific immune cell populations, and the CD nomenclature aids in this process by providing a uniform and unambiguous way to designate molecules of interest.

Attack of the Clones

Both T-independent and T-dependent activation lead to the rapid multiplication or "clonal expansion" of the responding B cell (figure 5.2). This is an essential first step in humoral immunity (and also in cell-mediated immunity, as we shall see later) since the starting number of naïve B cells with the ability to recognize any individual antigen is too small to be of any practical use. The proliferation of a single activated B cell produces a clone of about five thousand cells over the course of 5–7 days, with each cell of the clone expressing the same BCR and therefore capable of recognizing the same antigen that initiated the process.

A complex stimulus like a *Staphylococcus* bacterium is made up of multiple antigens, and so it will activate a large number of naïve B cells, each directed against a different antigen or a different part of an

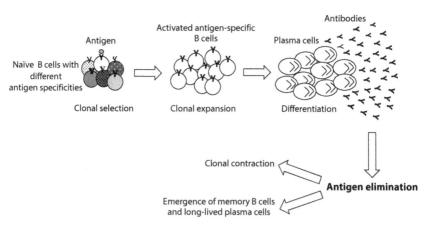

FIGURE 5.2. Phases of the B lymphocyte response. For simplicity, the activation of only a single antigen-specific B cell is shown.

antigen. Both T-independent and T-dependent mechanisms are likely to be involved since the bacteria have antigens that allow both modes of B cell activation. The proliferation of each of the responders will generate multiple clones of activated and numerically expanded B lymphocytes, resulting ultimately in thousands and thousands of cells with the ability to recognize the bacteria. Following the phase of clonal expansion, all the *Staphylococcus*-specific B cells differentiate into plasma cells, which are essentially cellular factories for churning out and secreting soluble forms of the BCR, otherwise known as immunoglobulins or antibodies. Some plasma cells differentiate relatively rapidly from the activated B cells, providing an early source of antibodies that help in initial elimination of the staphylococcal antigens, and then die a short time thereafter. Other plasma cells are generated only after the B cells have gone through the processes of isotype switching and somatic hypermutation (discussed in the next section).

The antibodies that are secreted by plasma cells enter the circulation from their site of production, the regional lymph node if the infection is localized or the spleen if the infection is blood-borne and systemic. They make their way to the infected area—the left thumb in the case of our patient—where they are able to get into the inflamed tissue because of the leakiness of the local blood vessels. Antibodies are the effectors of humoral immunity, the molecules that actually do the hard work of getting rid of antigens, including those associated with microbial pathogens like *Staphylococcus*. Since each plasma cell can produce about two hundred million antibody molecules per day, doing the math will tell you that we now have the numbers to take on the microbial invaders. Given the time taken to reach this point (about two weeks), you can also appreciate how important innate immunity is during the initial hours and days of an infection.

Bait and Switch

Antibodies come in four main flavors or isotypes. The isotypes are distinguished by relatively minor variations in the amino acid sequence of the constant region of the heavy-chain protein, small structural changes that can have a big impact because they confer different functional capabilities. The first isotype to be produced during the humoral response is called IgM, and the structure of its heavy chain corresponds to that of the BCR expressed by naïve B cells. The gene segment encoding the constant region of IgM is designated Cμ. IgM antibodies are produced following both T-independent and T-dependent B cell activation, generally by the short-lived plasma cells that differentiate early from the activated B cells. The other antibody isotypes— IgG, IgA, and IgE—are produced later in the response and mainly as a consequence of T-dependent activation. In humans, the IgG and IgA isotypes each have additional variants or subclasses—IgGs 1 through 4 and IgAs 1 and 2—that differ slightly in heavy-chain constant region structure. The corresponding gene segments encoding the constant regions of these other isotypes are designated Cγ1 to Cγ4, Cα1 and Cα2, and Cε. The various isotypes and subclasses are generated during B cell activation and proliferation by another somatic recombination process called isotype switching, which mainly takes place in specialized regions of the spleen and lymph nodes known as germinal centers (although some switching can also occur outside these areas).

In isotype switching, which involves another form of somatic DNA recombination, the rearranged VDJ gene segment encoding the variable region of the BCR IgM heavy chain of an individual B cell is grafted onto the C gene segment encoding the constant region corresponding to the IgG, IgA, or IgE isotypes. The process is controlled by special regions of DNA known as switch (S) regions, which are adjacent to each of the different constant gene segments. In the example of isotype switching shown in figure 5.3, the rearranged VDJ region is recombined so that it goes from being linked to the Cμ gene segment to the Cε segment, with the intervening DNA being looped out and removed. Following RNA processing and translation, the B cell now produces IgE heavy chains instead of IgM heavy chains. The IgE heavy chain can assemble with the existing light chain to form an IgE-type BCR. Since

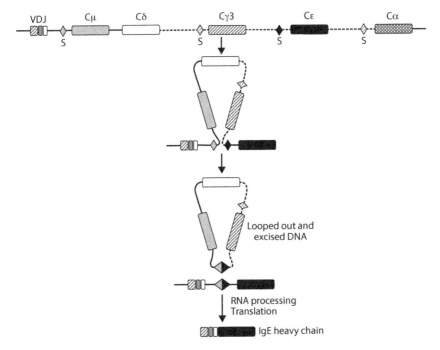

FIGURE 5.3. Isotype switching at the immunoglobulin heavy-chain locus, demonstrating how a B cell switches from IgM to IgE production while retaining the same antigen-binding specificity encoded by the rearranged VDJ segments. The Cγ1, Cγ2, and Cγ4 regions have not been shown for the sake of simplicity.

there has been no change in the VDJ regions, the antigen specificity of the BCR remains the same. Following activation, the IgE-expressing B cell will differentiate into plasma cells that secrete IgE antibodies with the same antigen binding ability as the originally produced IgM.

Isotype switching leads ultimately to the production of a pool of antibodies that have the same variable region (and, therefore, the same antigen specificity) but different constant regions. Since the constant region plays an important role in determining what kinds of functions the antibody is able to carry out, isotype switching diversifies the functional capabilities of the humoral response beyond those conferred by just IgM. The C gene segments that are chosen as targets during isotype switching, and therefore the relative proportions of the different antibody isotypes that are produced, are influenced by the specific cytokines and other factors present in the tissue microenvironment in which T-dependent B cell activation occurs. This contextual modulation of the process allows the generation of isotypes that vary with, and are generally most appropriate for, the specific antigenic stimulus that is involved. Bacterial infections, for instance, lead to an antibody response that is dominated by specific subclasses of IgG, whereas parasitic infections induce a lot of IgE.

Good Mutants

Besides isotype switching, T-dependent B cell activation is characterized by another type of antibody diversification mechanism, which also occurs in the rapidly proliferating B cells in the germinal center and is known as somatic hypermutation. As its name suggests, the process involves the introduction of multiple point mutations, specifically in the rearranged DNA encoding the variable regions of the heavy and light chains of the BCR. Since the heavy- and light-chain variable regions are the parts of the antibody that bind antigen, some of the introduced mutations will alter the tightness or affinity of binding to the target antigen. Somatic hypermutation thus leads to the generation of families of related B cells, with the members of each family originating from a single progenitor but having BCRs that differ in affinity for the same antigen. Why is this process of diversification important?

In the competition for binding to the limited amount of antigen present in the environment of the germinal center, B lymphocytes bearing mutated BCRs with high affinity for the antigen will bind more readily to the antigen than cells with mutated BCRs that have low or no affinity. The preferential binding of antigen to the former cells delivers signals that promote their survival and further proliferation, while the failure of the latter cells to capture antigen results in their elimination since they do not receive the necessary survival signals. As a result of repeated rounds of mutation, competition for antigen, and proliferation, B cells with higher and higher affinity for antigen increase in number at the expense of those with lower affinity. It's the age-old story of the survival of the fittest. Fortunately, this is a good thing in the case of germinal center B cells: the competition for antigen leads to a progressive increase in the affinity of the antibodies produced by the plasma cells that ultimately differentiate from the surviving B lymphocytes. This phenomenon is known as affinity maturation of the antibody response. Greater antibody affinity usually translates into more efficient elimination of antigen, so somatic hypermutation helps significantly to increase the efficacy of humoral immunity.

How Antibodies Get the Job Done

Antibodies clear foreign antigens and protect against microbial invaders through multiple mechanisms or effector functions, with different isotypes having different functional capabilities. The main antibody effector functions (figure 5.4) and the isotypes that typically execute them include neutralization (IgM, IgG, IgA), complement activation (IgM, IgG), opsonization (IgG), and antibody-dependent cell-mediated cytotoxicity (ADCC, IgG). In addition, IgG is involved in protection of the fetus and young infant, IgA in protection of mucosal tissues, and IgE in mast cell activation. Each of these functions is described below.

Neutralization

IgM, IgG, and IgA all have the potential to perform a function known as neutralization, which is the ability of an antibody to bind to a specific site on a microorganism or microbial molecule in a way that prevents the pathologic effects of those entities. Neutralization depends on where and how exactly the antibody binds: the binding must interfere with the activity of a microbial molecule that is crucial to the infection or disease-causing process. This function is particularly important in viral infections like COVID-19. The binding of neutralizing antibodies to the SARS-CoV-2 spike, a viral protein essential for cell entry, effectively prevents viral replication since the virus has

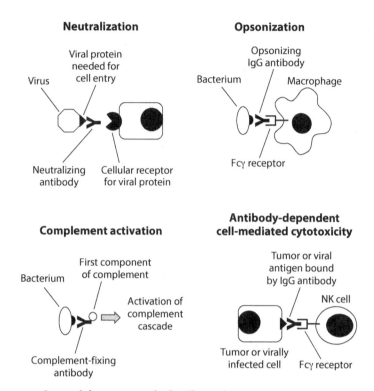

Neutralization

Virus

Viral protein
needed for
cell entry

Neutralizing
antibody

Cellular receptor
for viral protein

Opsonization

Opsonizing
IgG antibody

Bacterium

Macrophage

Fcγ receptor

Complement activation

Bacterium

First component
of complement

Activation of
complement
cascade

Complement-fixing
antibody

**Antibody-dependent
cell-mediated cytotoxicity**

Tumor or viral
antigen bound
by IgG antibody

NK cell

Tumor or virally
infected cell

Fcγ receptor

FIGURE 5.4. Some of the major antibody effector functions.

to get into a cell in order to multiply. Neutralization is also key to protection against diseases caused by bacterial toxins, which generally act by interacting with specific receptors on host cells. If binding of an antibody to the toxin prevents this interaction, the harmful effects of the toxin can no longer occur. One of the first therapeutic applications of antibodies was to neutralize the devastating pathological consequences of the diphtheria and tetanus toxins, which previously caused the deaths of hundreds of thousands of people.

Complement Activation

Another important antibody effector function is complement activation, which is carried out by the IgM and IgG isotypes. When IgM or IgG binds to its target antigen, the configuration of the antibody molecule is subtly altered, a relatively minor change that nevertheless leads to activation of the complement cascade. As we learned earlier, activated complement can help to eliminate microbial pathogens by facilitating their phagocytosis by macrophages and neutrophils (a mechanism that is also relevant to the clearance of any foreign antigen), by amplifying inflammation and by direct killing of certain bacteria. The activation of complement by antibody is a good illustration of the way in which the adaptive immune response cooperates with components of innate immunity to combat infecting microorganisms.

Opsonization

IgG, the dominant isotype produced during most immune responses, can act as a bridge between the target antigen, which is bound by the variable region of the antibody, and phagocytic cells, which latch on to the constant region through specific surface proteins known as Fcγ receptors. In this fashion, IgG that is bound to the surface of particulate structures such as bacteria makes it easier for macrophages and neutrophils to phagocytose the particle, an action similar to the opsonizing effect of complement. The triggering of Fcγ receptors that occurs during this process may have the additional effect of inducing the macrophage or neutrophil to deploy antimicrobial mechanisms or secrete various cytokines.

Antibody-Dependent Cell-Mediated Cytotoxicity (ADCC)

The binding of IgG to viral antigens on the surface of infected cells promotes the formation of Fcγ receptor-mediated interactions with NK cells. The NK cells are activated by this interaction to release molecules that kill the virally infected targets, an outcome that gives rise to the rather unwieldy name of this antibody effector function. ADCC is also effective against malignant cells, which are often characterized by the surface expression of abnormal proteins that are recognized as being nonself and that induce the production of IgG antibodies.

Both opsonization and ADCC depend on the binding of the constant region of IgG to Fcγ receptors. The constant regions of IgG1, IgG2, IgG3, and IgG4 have varying affinities for different Fcγ receptors (of which there are eight types, including activating and inhibitory subsets, expressed on different immune cells). Because of these differences, considerable diversity of IgG effector function can be achieved based on the relative amounts of the different subclasses.

Protection of the Fetus and Infant

One of the unique characteristics of IgG is its ability to cross from the maternal circulation into the fetus during the later stages of pregnancy. The transport of IgG across the placenta is carried out by another type of receptor known as FcRn. FcRn, which is expressed on placental cells, binds to the constant region of IgG and helps to ferry it from the mother's blood into the fetal circulation. This movement of IgG across the placenta allows maternal humoral defenses to be transferred to the fetus and provides crucial antimicrobial protection to the young infant during the first three to six months of life, a period that is particularly vulnerable to infection.

Mucosal Protection

IgA is the second most abundant antibody in the serum and is the predominant, almost exclusive, form found in the secretions of the gastrointestinal, respiratory, and genitourinary tracts. IgA plays a relatively minor role in the serum and systemic tissues, and the secretory form of the antibody is more important functionally, particularly in relation to interactions with the microbiota. It is produced in a dimeric form (a version in which two identical IgA molecules are linked together) by plasma cells located in mucosal tissues and is immediately grabbed by the poly-immunoglobulin receptor, a surface protein on epithelial cells. The dimeric IgA bound to the poly-immunoglobulin receptor is shuttled across the epithelium and then released on the other side into the lumen. This translocation of IgA across the epithelium of the gastrointestinal, respiratory, and genito-urinary tracts allows the antibodies to act preemptively against microorganisms in the lumen of the tissues, even before they get close to the epithelial cells.

Appropriately, isotype switching to IgA is promoted by molecules that are enriched in the environment of mucosal tissues, including metabolites that are generated with the help of the microbiota. In addition, some of the IgA that is produced at these sites is induced by antigens derived from certain members of the local microbial community. After being transported across the epithelium, this IgA binds to the organisms expressing the relevant antigens and keeps them under control, another example of the self-regulating nature of microbiota-host interactions. Secretory IgA can also be produced against pathogens that colonize or invade mucosal tissues, and it contributes to resistance against such organisms by neutralizing their ability to interact with the epithelium or by counteracting their virulence properties in other ways.

Mast Cell Activation

The IgE isotype is produced at very low levels in most people. In certain predisposed individuals, however, it can be expressed in excessive amounts as part of an aberrant response to normally innocuous food or environmental antigens. Most of the IgE is quickly and tightly bound by a specific receptor on the surface of mast cells, specialized innate cells that are located in the skin, mucosal tissues, and other sites, and that have a number of functionally potent proteins and lipids stored in cytoplasmic granules (figure 5.5). If the relevant food or environmental antigen is subsequently encountered, its binding to the IgE that is captured on the surface of mast cells triggers a concerted release of the granule contents. The substances released from the granules induce dilation and leakiness of local blood vessels, and also promote mucus secretion and muscle contraction in the gut and airways. These events manifest as the unpleasant and sometimes

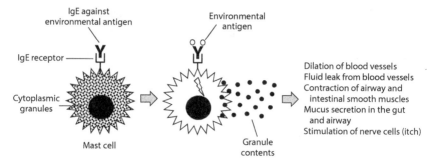

FIGURE 5.5. Activation of mast cells by IgE.

dangerous symptoms that are typically associated with allergies: hives, abdominal cramps, diarrhea and vomiting, difficulty breathing, and in the most severe cases, circulatory collapse and death.

It is not entirely clear why these IgE-mediated responses have evolved. They do not seem to confer any obvious benefits and, in fact, they may cause significant harm. The usual explanation put forward is that activation of mast cells by IgE can provide protection against some intestinal parasites: the increased production of mucus and the vigorous intestinal contractions—the "weep and sweep" reaction—can help in the elimination of such pathogens. There is also evidence that IgE may aid in resisting the action of some insect and reptile venoms. So, it is possible that sometime in our evolutionary past the ability to make IgE may have allowed our ancestors to survive in an environment where both parasites and poisonous animals were abundant. But why this vestigial protective response is inappropriately initiated in modern humans by antigens that pose no threat whatsoever remains a bit of a head-scratcher. This is a question that we will try to address in later chapters.

There is a fifth antibody isotype known as IgD. It is expressed mainly as a membrane-bound form on naïve B cells, with very small amounts being secreted. The functional significance of IgD antibodies is unclear. Some findings suggest that they play a role in interactions with respiratory bacteria, both components of the local microbiota and pathogens.

Antibodies Save the Day

Returning to what is going on in the little boy with the thumb injury, we can assume that the carbohydrate, lipid, and protein antigens of the infecting *S. aureus* will activate naïve B lymphocytes in the lymph nodes of the left armpit by T-independent and T-dependent mechanisms. IgM antibodies will be produced rapidly by the short-lived plasma cells that result from both modes of activation. The T-dependent component of the B cell response will subsequently lead to the production of isotype-switched, high-affinity antibodies. Under the influence of the kinds of cytokines that are typically secreted in response to PRR activation by the bacteria, the predominant isotype produced will be IgG. The IgM and IgG antibodies will make their way via the circulation to the infected thumb, and both types will bind to the bacteria and activate complement. The deposition of activated complement components on the bacterial surface, as well as the bound IgG itself, will facilitate phagocytosis and killing of the opsonized organisms. At the same time, other activated complement components will amplify the inflammatory response, bringing in more neutrophils and other immune cells to provide additional antimicrobial firepower. The collaborative interactions between the humoral response and innate immunity exemplified by these events, together with a little help from T cells, will lead ultimately to the complete eradication of the pathogen and a return to good health.

Defending the Future

Once the infection has been cleared successfully, most of the large numbers of activated B lymphocytes and short-lived plasma cells that were generated during the humoral response undergo programmed cell death. This phase of "clonal contraction" eliminates cells that have outlived their usefulness and that would otherwise consume valuable resources. However, a small fraction of the activated B cells—still expressing BCRs against staphylococcal antigens, usually in an isotype-switched and hypermutated form—are reprogrammed to become memory B lymphocytes. The reprogramming involves extensive changes in gene expression, as well as chemical modifications of DNA and chromatin proteins. These alterations confer a number of new characteristics on the cell, including an extended life span, fifty years or more in some cases, and the ability to respond more quickly and more robustly to antigen. The memory B cells circulate continuously through the blood and peripheral tissues so that if the same strain of *S. aureus* is ever encountered again, the humoral response will occur more rapidly and with greater magnitude and, hopefully, get rid of the pathogen before it can do any damage.

Along with the generation of memory B cells, some of the antibody-secreting plasma cells produced during the initial response are also reprogrammed to survive for many years, especially those that differentiate following isotype switching and somatic hypermutation in the germinal center. These long-lived plasma cells take up residence in the bone marrow and continuously secrete antibodies against staphylococcal antigens. The presence of this preformed antibody in the circulation provides an additional layer of protection against a second infection with the same organism. Like isotype switching and somatic hypermutation, the memory component of the humoral response, including the production of both memory B cells and long-lived plasma cells, is facilitated by T-dependent B cell activation (although some T-independent antigens *can* induce the development of humoral memory to varying extents).

We have covered a good deal of ground: the modes of B cell activation, the different antibody isotypes and their functions, and the generation of memory B lymphocytes and long-lived plasma cells. It is time well spent given how important B cells and antibodies are. They play essential roles in protecting us against infection and, as we will see in later chapters, their functions can be put to practical use in a variety of ways. But the humoral arm of adaptive immunity does have its limitations. Moreover, the B cell response itself would be much less effective if it did not have the assistance of T cells. T-dependent activation of B lymphocytes, with all the attendant benefits of isotype switching, affinity maturation, and the efficient induction of memory, would not be possible without the participation of T cells: it takes two to tango, as they say. For all these reasons, adaptive immunity always involves the activation of both B and T lymphocytes, and so it is to the latter cell type that we will now turn our attention.

CHAPTER 6

T Lymphocytes

A Little Help for My Friends

Antibodies are great at recognizing and dealing with antigens that are present in or exposed to the extracellular environment. Consequently, bacterial and fungal organisms that reside predominantly in this location are susceptible to the various antibody-mediated effector mechanisms discussed in the last chapter. Antibodies can also neutralize viruses prior to their entry into cells and can help to kill virally infected cells by means of ADCC. But many pathogens, including *M. leprae*, the causative agent of leprosy, as well as several types of bacteria and some protozoa, spend significant portions of their life cycle lurking *inside* cells. Antibodies, being large molecules that cannot cross the plasma membrane, are usually not very good at eliminating such organisms. Case in point: individuals with the lepromatous form of leprosy have plenty of antibodies directed against *M. leprae*, but they do very little to control the growth of the organism within macrophages. Because of this weakness of antibody-mediated defense, it helps to have an immune function that is capable of identifying and disposing of pathogens and other sources of foreign antigens that originate from or take refuge inside host cells. This is where T lymphocytes have important roles to play.

Antigen Presentation: Proteins In, Peptides Out

Like their B cell counterparts, newly formed naïve T cells circulate through the bloodstream to patrol peripheral lymphoid organs for the presence of foreign antigen. However, there are marked differences in the type of antigen recognized and the mode of activation between the two types of lymphocytes. In contrast to B cells, which can respond to proteins, lipids, carbohydrates, and even man-made molecules that are not found in nature, the majority of T lymphocytes are activated only by protein antigens (minor subsets of T cells *can* recognize other types of molecules, but we will not discuss them further since their functions are restricted to very specific types of responses). Moreover, the antigenic proteins have to be processed—broken down into short peptide fragments and the fragments displayed on the surface of another cell—before they can interact with and stimulate the TCR. These requirements mean that an antigen-presenting cell (APC) must participate in T cell activation.

While most cells can carry out restricted forms of antigen processing and presentation, the so-named professional APCs are dendritic cells, macrophages, and B cells. Dendritic cells, also sometimes known as veiled cells because of their diaphanous membrane protrusions or dendrites, are the most potent APCs and are generally thought to be the only ones capable of activating a naïve T cell. They are highly phagocytic in nature and constantly probe the environment for anything that can be ingested. Found in most tissues, they pick up molecules dissolved in the extracellular fluid, as well as particulates such as intact and dead microorganisms, and transport the collected material to the local lymph nodes by migrating into and through lymph vessels. This migratory activity occurs at a low level under basal conditions but is enhanced in the presence of PAMPs or DAMPs. The material that is phagocytosed and endocytosed by the dendritic cell is delivered to an intracellular compartment and is broken down by enzymatic action, generating degradation products that usually include short peptide fragments of proteins. It is these peptides that are presented to T cells for recognition as antigens. However, in order for such presentation to occur, the peptides must be taken to the cell surface in some way. This important transport function is carried out by class II MHC molecules.

Each class II MHC molecule consists of an α chain and a β chain that are assembled into a complex (figure 6.1, *left* and *middle*). One end of

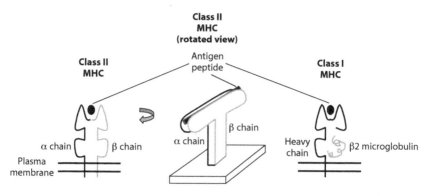

FIGURE 6.1. The class I and class II MHC molecules. Both types of molecules have similar overall structures, as illustrated by the rotated view of the class II MHC.

the molecule is embedded in the cell membrane, while the other end, which points toward the exterior of the cell, forms a shape that looks a lot like a hot dog bun. On the way to the plasma membrane, the assembled α and β chains enter a compartment where they encounter the peptides that have been produced by degradation of endocytosed and phagocytosed substances. Individual peptides that have the appropriate structure are loaded into the grooved end of the molecule, very much like a hot dog being slotted snugly into its bun. Each dendritic cell, like other professional APCs, expresses six different class II MHC molecules, which vary in the fine structure of their grooves so that together they can accommodate a wide range of peptides.

The loaded MHC molecule now proceeds to the cell surface, where it displays its peptide cargo to any T cell that might be passing along. T lymphocytes with the "wrong" TCRs will keep moving, perhaps after a brief pause to check out the dendritic cell displaying the peptide-loaded MHC. But if the encounter involves a T cell with a TCR that has affinity for the *combined* structure of the particular peptide and specific MHC molecule, it will stop for a prolonged conversation with the dendritic cell that will determine the subsequent course of the T cell response. (Sounds a bit like speed dating, doesn't it?) Similar events occur when the APC happens to be a macrophage or a B cell, although usually only if the T cell has already been activated by interacting with a dendritic cell. In fact, we saw in the last chapter that this type of MHC-mediated interaction is involved in the T-dependent activation of B cells.

The class II MHC pathway is dedicated to the processing and presentation of *exogenous* proteins, those that are taken up from the extracellular environment (figure 6.2, *left*). But foreign antigens can also originate within the cell, as in the case of viral infection, where the virus uses the host cell's biosynthetic machinery to make the proteins needed for its life cycle. In addition, neoplastic (tumor) cells sometimes express proteins that are effectively nonself or foreign because they are not made by normal cells. Such *endogenous* viral or tumor-associated proteins must also be processed and presented in order for T cells to recognize and respond to them. This function is carried out by class I MHC.

The class I MHC molecule consists of a heavy chain (not to be confused with the immunoglobulin heavy chain) that is embedded in the

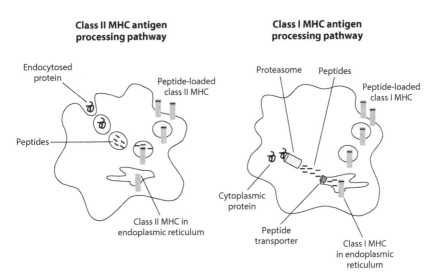

FIGURE 6.2. Pathways of antigen processing and presentation by class I and class II MHC molecules.

membrane and that is associated with a soluble protein called β2 microglobulin (see figure 6.1, *right*). The heavy chain forms a structure that is very similar to that of its class II relative, including a grooved end into which peptides can be loaded. Class I MHC also comes in six varieties per cell so that a large number of different peptides, almost all derived from the breakdown of endogenously synthesized proteins, can fit into one or the other of them. The way in which class I MHC acquires its peptide is a little different from class II (see figure 6.2, *right*). Endogenous proteins in the cytoplasm, including self proteins that are misfolded or that have reached the end of their functional life, as well as proteins that are generated as a result of viral infection or neoplastic transformation, are degraded into peptides by a large enzymatic complex known as the proteasome. The proteasome functions like a wood chipper: proteins are fed in at one end and short peptides are ejected at the other. These peptides are threaded through a special transport channel in the membrane of the endoplasmic reticulum and loaded onto newly synthesized class I MHC molecules that are being readied for movement to the plasma membrane. After picking up its peptide cargo, the class I MHC molecule migrates to the cell surface, where it is available for recognition by a T cell with the appropriate TCR. While the class II pathway of antigen presentation is largely restricted to professional APCs, the class I pathway operates in all nucleated cell types since virtually any cell can be virally infected or give rise to a tumor.

Although we have discussed the class I and class II MHC pathways as if they were strictly segregated, they do intersect, at least in some cell types. Dendritic cells in particular have the ability to load peptides that are generated from ingested, exogenous protein antigens onto the class I MHC molecule. This process, known as cross-presentation, is crucial to the ability of these cells to present tumor-associated and viral peptides on class I MHC, an essential step in initiating a T cell response to tumors and viruses if the dendritic cells themselves are not neoplastic or infected by the virus.

The Need for MHC Molecules

The advantage of having TCRs recognize the combined structure of a peptide loaded onto an MHC molecule is not immediately obvious, but there are a couple of ideas that are worth mentioning. The first is that the breakdown of proteins may be necessary to reveal foreign antigenic determinants that might otherwise be hidden within the folds of the three-dimensional structure of the molecule. Once a protein is broken down, the resulting peptides have to be transported in some way to the cell surface in order to be available for recognition by the TCR. That requirement is fulfilled by the class I and class II MHC molecules. The second idea is that having the peptide antigen bound to the MHC molecule on the surface of a cell forces the T lymphocyte to focus its attention on the source of the antigen instead of being "distracted" by free-floating antigenic molecules (which, in any case, can be dealt with by antibodies). This focus is likely to be particularly important in viral infections, where it is probably more efficient to try to eliminate the virally infected cell rather than all the viral particles released by that cell.

Regardless of whether these speculations are correct, the requirement for the TCR to recognize a complex of peptide and MHC leads to a unique and important step of development that T cells, but not B cells, have to go through. We have seen earlier that developing B and T lymphocytes are subjected to a process of negative selection in which cells with self-reactive antigen receptors are eliminated. In the case of developing T lymphocytes, an additional mechanism known as positive selection allows only those cells that bear TCRs with a modest degree of affinity for the grooved, hot dog bun end of self MHC molecules to survive and develop further. Positive selection ensures that the mature T cells that emerge from the thymus are predisposed to recognizing MHC-like shapes and are thus more likely to be triggered by foreign peptides presented by self MHC molecules. If the affinity of the TCR for self MHC is too high, however, the developing T cell will be induced to undergo negative selection.

The MHC and Me

A cell typically expresses six types of class I MHC molecules, half of them encoded by three separate genes on a chromosome inherited from the mother and the other half by three genes on the corresponding chromosome from the father. If the cell happens to be a professional APC, like a dendritic cell, macrophage, or B lymphocyte, it will also have six types of class II MHC molecules, again with three encoded on each parental chromosome. Each type of MHC molecule differs from the others in the amino acids that make up the sides and floor of the groove. Between the six types of class I MHC molecules in a cell (and six types of class II MHC molecules if the cell is a professional APC), most antigenic peptides can be bound. Although the repertoire of MHC molecules expressed by an individual is limited to six each of the class I and class II varieties, there is much greater MHC diversity across the human *population*, more than a thousand variants or alleles for each of the corresponding genes. This population diversity helps to ensure that the whole human race cannot be wiped out because of an inability to present an antigenic peptide derived from a deadly pathogen (even if a significant proportion of individuals lacks that ability).

The diversity of MHC at the population level also means that the complement of MHC molecules expressed by individual A is likely to be very different from individual B unless the two happen to be identical twins. The interindividual difference in MHC repertoire can give rise to some variations in immune responsiveness between one person and another, usually to specific types of antigens. In most individuals and under most circumstances, such variation is relatively inconsequential and is well worth the evolutionary advantage of avoiding extinction of the species. However, the difference in MHC molecules between individuals does pose a problem for present-day humans, specifically, those who require an organ transplant. The immune system of individual A will inevitably recognize the MHC molecules of individual B as being foreign and will mount an anti-B response (and vice versa). MHC differences thus represent a major barrier to transplanting organs from one individual to another, a barrier that can be overcome only by suppressing immune responses in the recipient.

Turning T Cells On

The stage is now set for T cell activation (figure 6.3). The major players—dendritic cells displaying peptide-loaded MHC molecules on their surface and T cells expressing a variety of TCRs—have arrived at their places in the lymph node, and the action is about to begin. Two important subsets of T cells are involved, which are distinguished by the expression of the CD4 or CD8 surface proteins and are usually identified as being CD4+ or CD8+, respectively. CD4 is found on T cells bearing TCRs that recognize peptides presented by class II MHC, while CD8 marks T cells with TCRs that recognize peptides presented by class I MHC. If a T cell encounters a dendritic cell with a peptide-MHC combination that is recognized by its TCR, that T cell, out of all the others in the vicinity, is clonally selected for initiation of a stable interaction. The interaction is facilitated by the binding of CD4 or CD8 to class II or class I MHC, respectively, and ultimately results in the delivery of a signal through the TCR that immunologists refer to as Signal 1. But Signal 1 on its own is not sufficient to activate a naïve T lymphocyte. In fact, a T cell that receives only Signal 1 may become unresponsive to further stimulation, entering a zombie-like state known as anergy, or may even die. It is only when another signal, known as Signal 2, is delivered simultaneously with Signal 1 that the T lymphocyte becomes fully activated and enters the phase of rapid proliferation and clonal expansion. Signal 2 results from the interaction between a co-stimulatory molecule on the surface of the dendritic cell (an example is the protein called B7-1, but there are others) and a corresponding receptor on the surface of the T cell (CD28, for example). Dendritic cells do not normally express co-stimulatory molecules and are induced to do so only if they have been activated previously through a PRR by a PAMP or DAMP.

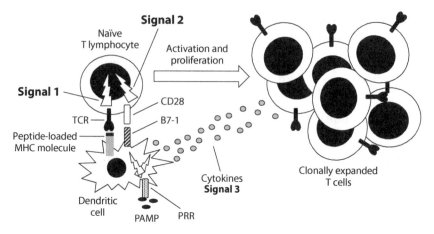

FIGURE 6.3. Activation of a naïve T lymphocyte by an antigen-presenting dendritic cell.

On the face of it, the dependence of T cell activation on two contemporaneous signals—one sparked by the binding of peptide-MHC to the TCR and the other by a dendritic cell co-stimulatory molecule (which itself has to be induced by PRR activation)—seems unnecessarily complicated, almost baroque in its intricacy. But there is method in this apparent madness. The requirement for the two signals reduces the likelihood that T cells will be activated by antigen in the absence of a microbial or tissue-damaging threat. This is a particularly important consideration since dendritic cells and other APCs do not discriminate between self and foreign antigens. They will degrade any protein that is delivered to their antigen processing machinery and will present any of the resulting peptides that can be loaded onto either the class I or class II MHC molecules. That being the case, they could activate one or more of the handful of self-reactive T cells that inevitably escape negative selection in the thymus if only Signal 1 was required. In actuality, the delivery of Signal 1 without Signal 2, something that is more likely to occur under basal conditions, when most of the peptides presented by dendritic cells are derived from self proteins and PAMPs and DAMPs are lacking, *prevents* self-reactive T cell clones from responding and acts as a safeguard against the development of autoimmune problems. When Signal 2 is triggered along with Signal 1, it indicates to the T cell that circumstances are such—"Microbes have invaded! Tissues are damaged! Foreign antigens are abundant!"—that a response is warranted.

In addition to expressing co-stimulatory molecules, dendritic cells respond to PAMPs and DAMPs by secreting various cytokines. Some of these molecules have a significant influence on the functional differentiation of the T lymphocytes that are being activated by Signals 1 and 2, and for that reason they are sometimes viewed as a third signal, Signal 3, that regulates T cell activation.

Lifting the Veil on Veiled Cells

It should be obvious by now that dendritic cells play a crucial role in activating T cells: they ferry antigens from peripheral tissues to local lymph nodes, they degrade the antigens and present the resulting peptides on MHC molecules to deliver Signal 1, and they respond to PAMPs and DAMPs by expressing the co-stimulatory molecules that deliver Signal 2. The co-stimulatory function provides an essential link between innate and adaptive immunity, ensuring that the two sets of responses are coordinated and, importantly, that the latter does not occur without the former. The dependence of adaptive immunity on the activation of innate immunity is the basis for what the late Charles Janeway, one of the doyens of modern immunology, called "the immunologist's dirty little secret," namely, that you cannot elicit good B and T cell responses to an antigen unless the antigen is "contaminated" with a bit of microbial material that provides a PAMP. That is why antigens used to elicit B or T cell responses in mice are usually mixed with something called Freund's adjuvant, a witch's brew of different substances, including PAMPs derived from mycobacteria. Human vaccines have to include some type of adjuvant for the same reason.

Despite their importance, dendritic cells were unheard of until the 1970s, when Ralph Steinman, a physician-scientist at the Rockefeller University in New York, first identified them as a distinct cell type. His studies laid the foundation for a whole new area of research, one that has revealed the multiple ways in which the cells participate in and control the immune response, and that continues to flourish today. His contributions to this work led to the 2011 Nobel Prize in Physiology or Medicine that, tragically, was announced three days after he died of pancreatic cancer. The prize is generally not awarded posthumously, but the Nobel committee decided to make an exception in this instance since it was unaware of his death when he was selected for the honor. In a fitting turn of events, which some might consider the workings of karma, the cells that Steinman brought to the scientific world's attention were at least partly responsible for extending his life beyond the six months that he had been given when the cancer was first diagnosed. A determined researcher to the end, he carried out his final, and certainly most poignant, experiments on himself, using dendritic cells in a series of treatment strategies that he devised and that may have helped to slow the progression of his tumor for some time. He must have had great faith in the cells that he had stumbled on forty years earlier, and in his understanding of their crucial functions.

T Cells Take Action

To appreciate how T cells contribute to the adaptive immune response, let's take a look at what happens when an individual becomes infected by the influenza virus, something that occurs all too frequently during flu season. The virus is transmitted through aerosolized droplets that are expelled when someone with the infection coughs or sneezes. It is inhaled into the respiratory tract of another person who has the misfortune of being nearby, enters the epithelial cells of the air passages and air sacs, and starts to multiply. Viral nucleic acids, the PAMPs in this scenario, are recognized by PRRs on the epithelial cells themselves, as well as on lung immune cells such as macrophages and dendritic cells. The interaction of PRRs with the viral PAMPs, and with DAMPs released from damaged cells, induces the cells to secrete type 1 interferons and various cytokines, chemokines, and eicosanoids.

The interferons act on surrounding cells to enhance their resistance to viral infection, while the cytokines and other mediators activate an inflammatory response, including the influx of neutrophils and other cells from the local circulation. The lung inflammation gives rise to the typical symptoms of cough, difficulty breathing, and fever.

While all this is going on, dendritic cells have ingested the virus, either free-floating or contained within infected, dead epithelial cells, and have transported the viral proteins into the class II MHC pathway and also, as a result of cross-presentation, into the class I MHC pathway. Activated and made more motile by stimulation of their PRRs, the dendritic cells enter the lung lymphatic vessels and migrate to the local lymph nodes, which are located in the middle of the chest cavity. By the time the cells reach their destination, they have increased their expression of co-stimulatory molecules and the viral proteins they have taken up have been degraded into peptides. The peptides are displayed on the cell surface bound to either the class I or class II MHC molecules and are available for recognition by the appropriate CD8+ or CD4+ T lymphocytes, respectively. The response of both types of T cells follows a sequence of steps similar to the ones that occur in humoral immunity: clonal selection, clonal expansion, and differentiation of effector cells (figure 6.4). The details of how CD8+ T lymphocytes behave during the response are fairly straightforward, so we will start with a discussion of the functions of that subset.

Killer Ts

A CD8+ T cell expressing TCRs with the ability to bind to a specific combination of viral peptide and class I MHC molecule will interact with the presenting dendritic cell and receive Signal 1. It will also get Signal 2 since activated dendritic cells express the necessary co-stimulatory molecules. Having received both signals, the T cell is triggered to proliferate and undergo clonal expansion over the course of several days, generating a number that is sufficient to help eliminate the infection. During this process, the cells exit the lymph node

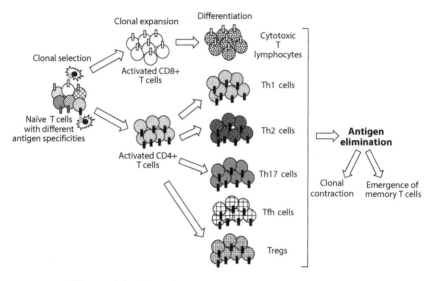

Clonal expansion

Differentiation

Clonal selection

Activated CD8+
T cells

Cytotoxic
T
lymphocytes

Naïve T cells
with different
antigen specificities

Th1 cells

Activated CD4+
T cells

Th2 cells

Antigen
elimination

Th17 cells

Clonal
contraction

Emergence of
memory T cells

Tfh cells

Tregs

FIGURE 6.4. Phases of the T lymphocyte response.

and travel via the bloodstream to the inflamed lung, where they leave the circulation and migrate into the tissue. The expanded clones of T cells, each clone directed against a different viral peptide antigen-class I MHC combination, ultimately differentiate into effector cells, either on their way to the lung or after reaching there and interacting with local innate cells.

Although CD8+ T cells are capable of carrying out several different effector functions, they are probably best known for their ability to kill virally infected or cancerous cells (figure 6.5, *top*), which is why they are often referred to as cytotoxic T lymphocytes, or CTLs. In order to execute their cytotoxic function, the CTLs must first identify their targets in the influenza patient, the virally infected cells within the lung tissue. They do so by searching for cells marked by the surface expression of the specific combination of viral peptide and class I MHC that originally activated them. Such cells will be enfolded by the CTL in a lethal embrace that is initiated by the binding of the peptide-MHC complex to the TCR. A kiss of death follows, delivered by an array of secreted and surface molecules expressed by the CTL, some that punch

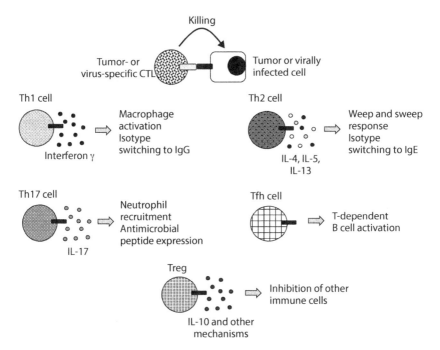

FIGURE 6.5. Effector T cell functions.

holes in the membrane of the infected cell, others that induce it to commit suicide. With its home wiped out, the virus no longer has a place to shelter and multiply, and the infection is contained. In addition to its death-dealing molecules, the CTL may also secrete various cytokines that amplify the inflammatory response and boost the antiviral functions of adjacent epithelial and innate immune cells.

Helpers and Regulators

As CD8+ T cells are being activated, CD4+ T cells in the lymph node that bear TCRs for specific combinations of viral peptides and class II MHC molecules will also receive Signals 1 and 2 from the antigen-presenting dendritic cells and will be induced to undergo clonal expansion. The expanded clones will leave the lymph node and migrate to the lung, differentiating into effector cells in the process. There are several types of effector CD4+ T cells (figure 6.5), and they are usually distinguished by the pattern of cytokines that they secrete. Some of them function by helping other cells of the immune system and are classified as T helper cells, or Th for short. Several subsets of Th cells have been identified, and each subset is specialized to carry out different functions.

Th1 cells are characterized by the secretion of the cytokine interferon γ, a functional relative of the interferons involved in innate immunity. Interferon γ is particularly effective at enhancing the antimicrobial activities of macrophages and other cells, and thus helps in dealing with pathogens that reside inside cells, including various viruses and certain types of bacteria like those that cause typhoid fever, tuberculosis, and leprosy. In the tuberculoid form of the latter disease, it is a robust Th1 response that helps to restrict the growth of *M. leprae* inside macrophages.

Th2 cells secrete IL-4, IL-5, and IL-13. These cytokines act on intestinal epithelial cells to increase mucus production, on macrophages to enhance certain types of antimicrobial mechanisms, and on B cells to induce switching to the production of IgE. As we learned in an earlier chapter, IgE triggers mast cells to release various bioactive molecules, including some that cause the intestinal musculature to contract. Both the increased mucus production and the vigorous muscle contractions facilitate the expulsion of intestinal parasites (the weep and sweep mechanism). Th2-activated macrophages may also aid in dealing with parasitic infections.

IL-17 produced by Th17 cells promotes recruitment of neutrophils and the secretion of various antimicrobial peptides by these cells and by the epithelial cells lining the intestine and other mucosal sites. These actions help in clearing extracellular pathogens, including strains of *Staphylococcus* and *Streptococcus* that cause bacterial pneumonias.

Finally, there is a specific Th cell type known as the T follicular helper cell or Tfh. By expressing CD40L and secreting cytokines such as IL-4 and IL-21 following activation, these cells play an important role in T-dependent B cell activation and in promoting the processes of isotype switching and somatic hypermutation. Other, more minor Th subsets, each with their own characteristic armamentarium of cytokines, may be found in the context of specific infections or disease states.

There is an additional type of CD4+ effector T cell that does not act as a helper but as a regulator of other immune cells. It is known as the T regulatory cell, or Treg (see figure 6.5, *bottom*), and it expresses secreted and surface molecules that it uses to inhibit the activity of Th cells, macrophages, and other cell types. The inhibitory functions of Tregs help to ensure that immune responses do not go out of control. They are particularly abundant at mucosal surfaces, especially the lower part of the intestine, where they prevent the development of excessive or inappropriate reactions to the microbiota. Like IgA, a significant fraction of the Tregs present in mucosal tissues are, in fact, induced by the microbiota and carry TCRs that recognize peptide antigens derived from the resident microbial community. This is one more example of how the microbiota helps to maintain peaceful coexistence with its host.

The type of effector that activated CD4+ T cells differentiate into is influenced by the cytokine environment in which antigen recognition occurs, including the cytokines that are secreted by the antigen-presenting dendritic cell and that represent Signal 3. The characteristics of Signal 3, in turn, are shaped by the kind of PAMPs and DAMPs that the dendritic cell is exposed to. Because of this cascading series of influences, the nature of the pathogen is an important factor in steering the differentiation of activated CD4+ T cells into specific effector types. In general, the cytokines induced by a pathogen direct the differentiation of effector T cells that are most suited to dealing with that category of organism. As a result, infections caused by viruses, including influenza, typically elicit a T lymphocyte response that is dominated by CD8+ CTLs and Th1-type CD4+ effectors. Intracellular bacterial pathogens also activate a Th1-type response, whereas parasites induce a Th2 response and extracellular pathogens a Th17 response.

Although it is conceptually convenient to think of activated CD4+ T cells differentiating into distinct effector subsets, it is important to mention that these cells actually have considerable functional plasticity and that the distinctions between subsets are sometimes fuzzy and not always set in stone. A T cell that starts out with the characteristics of the Th17 subset, for instance, may display features of Tregs during later stages of the response, or vice versa. This shape-shifting behavior may reflect adaptations to changes in local microenvironmental conditions and may allow T cells to respond more appropriately to ongoing variations in the status of the infection. But it does make it harder for immunologists to characterize and describe T cell responses, and it can be a source of some confusion and controversy.

Termination of T Cell Responses

The combined actions of CD8+ CTLs and CD4+ Th cells, along with the antiviral antibodies that are produced by activated B cells, will ultimately clear the influenza virus from our patient, probably after a few unpleasant days of fever, cough, and other symptoms. Once the viral antigens are completely eliminated, most of the T cells, as well as the B cells that responded to the virus, undergo programmed cell death and clonal contraction because they no longer receive activating signals. In addition to this passive "foot off the accelerator" mechanism, a more active "apply the brake" strategy also comes into play to ensure that baseline conditions are restored. The brake is applied as the result of specific inhibitory receptors that are expressed on the surface of activated T lymphocytes. These inhibitory molecules turn off T cell proliferation and effector functions following interactions with specific ligands expressed on surrounding cells. Several T cell inhibitory receptors have been described. Two of the more well studied ones, which we will encounter again in a later chapter, are PD-1 and CTLA4. In addition to the action of inhibitory receptors, Tregs and pro-resolving lipid mediators may also contribute to the dialing down of the T cell response.

Memory and Exhaustion

As in the case of the B cell response, a small number of the activated T cells persist as long-lived memory cells after the phase of clonal contraction. Three types of memory T cells are currently recognized: the central memory T cell, or T_{CM}, which circulates mainly through the blood and lymph nodes and is poised to differentiate into various effector types following activation; the effector memory T cell, or T_{EM}, which migrates through peripheral tissues and is pre-differentiated into specific effector types that are already equipped to deploy their protective functions once they have expanded in number; and the resident memory T cell, or T_{RM}, which is a nonmigratory cell that remains confined to certain tissues and is able to provide local protection. In all cases, the memory T cells continue to express the TCRs that were originally activated by viral peptides loaded on class I or class II MHC (TCRs, unlike BCRs, do not go through isotype switching or somatic hypermutation), but they have been reprogrammed to have an extended life span and to respond more quickly and robustly than naïve T cells if the same antigens are encountered again.

Some infections are not cleared efficiently by the adaptive immune response. The persistence of relatively high levels of microbial antigen in such circumstances is associated with chronic stimulation of T cells through their TCRs, which leads to the increased expression on these cells of several inhibitory receptors, including PD-1 and CTLA4. The interaction of the inhibitory receptors with their corresponding ligands on other cells in the vicinity, including macrophages and various non-immune cell types, pushes the previously activated T lymphocyte into an "exhausted," dysfunctional state that is characterized by impaired, although not completely absent, ability to proliferate, secrete cytokines, and execute cytotoxic activity. The exhausted state is typically associated with CD8+ T cells, but there is evidence that it also affects CD4+ T cells. T cell exhaustion may represent a regulatory mechanism that helps to prevent the tissue damage that could result from continued and excessive T cell activity, but it also limits the ability to clear infection, resulting in a host-pathogen stalemate in which neither side can claim outright victory. Exhausted T cells have been implicated in the chronicity of several human infectious diseases, including AIDS, hepatitis B, hepatitis C, and tuberculosis. They are also associated with several types of cancer. As we shall see later, blocking the action of the inhibitory receptors that contribute to the exhausted state has been used successfully for the treatment of certain tumors. Unfortunately, this strategy has been less effective for the management of chronic infections, at least so far.

What we've learned up to now has given us insight into the basic workings of the immune system, the broad mechanisms that mediate and govern its function. We have seen that epithelial defenses and relatively stereotypic innate immune responses provide broad protection during the early phase of infection while the more sophisticated machinery of adaptive immunity is being started up and set in motion. Once a sufficient number of B and T cells have been clonally activated and expanded, their diverse functional capabilities, tailored to the specific pathogen involved, act together with elements of the innate immune system to eradicate the infection. Following clearance of the infecting microorganism, activating signals subside and inhibitory ones come into play, clonal contraction of lymphocytes takes place, and the hot fires of inflammation are cooled. Memory lymphocytes and long-lived plasma cells emerge to take up their roles as patrollers and guardians, ensuring a more rapid and effective response if the same pathogen is seen in the future. The remainder of our discussion will focus on applying all this information—a summary of more than a century's worth of work—to the understanding and treatment of disease.

CHAPTER 7

Immune Dysfunction

When Things Go Wrong

Like Goldilocks's taste in porridge and beds, the functioning of the immune system has to be just right in order to maintain the balance between adequate protection against infection and avoidance of excessive and potentially harmful inflammation. Either too little or too much immune activity can lead to problems (figure 7.1)—the former to diseases known as immunodeficiencies, the latter to a diverse group of disorders characterized by inappropriate, unregulated, and damaging responses.

Too Little Immunity Part I: Acquired Immunodeficiencies

As the name suggests, an immunodeficiency represents a lack of one or more important immune functions. Since the immune system exists to protect against microbial invasion, immunodeficiencies generally manifest as recurrent and sometimes unusually severe infections, including by organisms that are not harmful under normal circumstances.

The more commonly seen types of immunodeficiency occur during or after certain infectious diseases, or in association with conditions such as malnutrition, diabetes, and cancer. One of the most studied and well-known examples of this type of immunodeficiency is the acquired immunodeficiency syndrome (AIDS), which is caused by infection with the human immunodeficiency virus (HIV). HIV directly infects, and ultimately destroys, CD4+ T cells. Based on what we know about the importance of these cells and their involvement in

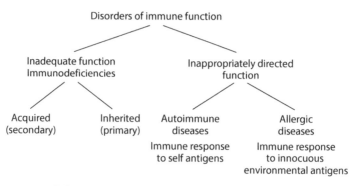

FIGURE 7.1. Broad classification of the disorders of immune function.

multiple aspects of the immune response, it is easy to understand how their loss can lead to widespread deficits of immunity. The progressive immunological impairments that result from untreated HIV infection lead to a plethora of symptoms, including frequent infections with bacterial and fungal pathogens and an increased risk of developing some types of cancer.

Infectious diseases other than AIDS can also compromise immune defenses, sometimes for periods of weeks to months after the infection. Sepsis, a potentially fatal malfunction of multiple organs that can occur during several kinds of infections, is often associated with a generalized suppression of immune function. The exact mechanisms that contribute to impaired immunity in sepsis are not clear. Measles is another example of postinfection immunodeficiency. Caused by a highly transmissible respiratory virus, measles usually affects children and manifests with fever, cough, and a characteristic skin rash. In most cases, the illness resolves over several days without causing further problems. But in the era before measles vaccination was widely implemented, it was not uncommon to encounter a young child who had recovered from measles only to develop a serious bacterial pneumonia soon afterward.

When I was training as a pediatrician in India, a particularly dreaded complication of measles was disseminated TB. Many children in developing countries are infected by *Mycobacterium tuberculosis*, the organism that causes TB, but their immune systems are able to contain, although not completely eliminate, the infection. The immune cells and the bacteria exist in a precarious equilibrium. Under normal circumstances, few, if any, symptoms or clinical abnormalities are apparent in this latent form of TB. However, if the immune system becomes weakened after measles, *M. tuberculosis* starts to get the upper hand. It grows and spreads, first to multiple areas of the lungs and then, in the worst cases, to the delicate membranes surrounding the brain and spinal cord. Typically, a child who seems to be recovering well from measles will gradually start to lose weight and become listless, irritable, and uninterested in food or play. A chest X-ray taken at this stage will often show innumerable, small foci of tuberculous infection scattered throughout both lungs, like snowflakes dotting the night sky. If the central nervous system is affected, the child may be brought to the emergency room with convulsions or in a coma. Unfortunately, death is often not far behind in such patients.

Until recently, we lacked a good explanation for the immunocompromised state that follows measles. However, research conducted over the last four to five years has revealed that the measles virus has a predilection for infecting and killing lymphocytes, particularly those that have differentiated into the cells that represent the repository of immunological experience—the memory B cells, memory T cells, and the antibody-producing long-lived plasma cells that survive following encounters with pathogens or vaccines. Essentially, measles wipes out immunological memory, inducing a condition that has been called "immune amnesia." This state of forgetfulness is particularly vulnerable to infections that would otherwise be prevented by preformed antibodies or cleared efficiently by memory lymphocytes.

Too Little Immunity Part II: Inherited Immunodeficiencies

Besides the impairments of immunity that occur secondary to infections or other health problems, there is a class of immunodeficiency disease that is the result of primary defects in the development or function of immune cells. If you think of the incredible complexity of the immune system, the multiple cell types and molecules involved, and the intricate ways in which these components have to interact in order to mount an effective response, it is easy to appreciate how the failure of even one part can prevent the whole from functioning correctly. The generation, characteristics, and behavior of all those cells and molecules require the coordinated operation of thousands and thousands of genes, each of which has to be transcribed and translated at the right time and in the right cells. Genetic mutations that significantly interfere with any of these processes can compromise the ability to sense or respond to infection. Such mutations are important, although rather unusual, causes of immunodeficiencies.

About four hundred single-gene defects that impair immune function have been described. Individually, most of them are quite rare, with some affecting only a handful of patients worldwide. They usually come to clinical attention when someone, usually a young child, suffers from repeated bouts of infection involving the skin, soft tissues, lungs, or other organ. The kind of pathogen and the severity of the infection are influenced by the specific aspect of immunity that is impaired by the mutation and can provide clues to the underlying abnormality. With the right tests and the relevant resources, it is often possible to arrive at a precise diagnosis of the problem, including the identity of the gene that is mutated.

We have already encountered one of the more frequently seen of these genetic disorders of immune function in an earlier chapter—CGD, which is the result of mutations in genes involved in generating the oxidative molecules that phagocytes use to kill bacteria. The lack of this important antimicrobial defense mechanism results in a predisposition to repeated bacterial infections of the lungs and soft tissues underneath the skin. Another relatively common genetically determined immunodeficiency is X-linked agammaglobulinemia (XLA). This disorder is caused by mutation of a gene known as BTK, which is required for the normal development of B lymphocytes. Individuals with XLA have abnormally low numbers or even complete absence of B cells, with correspondingly low levels of circulating antibodies and defective humoral immune responses. They suffer from recurrent bacterial infections of the lungs and sinuses, systemic viral infections, and certain parasitic infections of the gastrointestinal tract, reflecting the importance of antibodies in protection against these pathogens.

There are other, rarer genetic disorders of immunity that affect the development or function of various immune cell populations. One group of these diseases is associated with mutations that impair the development of both B and T cells, giving rise to the kind of severe combined immunodeficiency (SCID) that was made famous by the poignant story of the "boy in the bubble." The absence of the two key cell types of adaptive immunity made this young child so immunocompromised and so susceptible to a variety of pathogens that he had to spend the entirety of his brief life inside a plastic isolator in order to avoid infection. Mutations that impair the production or sensing of the cytokine interferon γ are associated with defects in the generation of an effective Th1 response and lead to a condition known as Mendelian susceptibility to mycobacterial disease (MSMD). As you might expect from our understanding of the role of Th1 responses, MSMD predisposes specifically to infections with bacterial pathogens like *M. tuberculosis* that reside within macrophages.

Recently, rare, inactivating mutations in genes required for the production of antiviral type 1 interferons have been identified in individuals who develop life-threatening SARS-CoV-2 infection. Interestingly, a follow-up study found that about 10% of people with severe COVID-19 had circulating antibodies that inhibited the function of these interferons. Although the exact role of these interferon neutralizing antibodies in the disease process remains to be determined, their presence in a significant number of seriously ill COVID-19 patients is suggestive and illustrates how the study of rare mutations can have relevance for the understanding of disease mechanisms in the general population. This idea holds true more broadly: although genetically determined immunodeficiencies affect relatively small numbers of people, investing time and resources into clarifying their molecular basis returns substantial dividends, including insights into the functions of individual genes in the human immune response and the possibility of using that information to develop treatment strategies.

The single-gene mutations that give rise to immunodeficiencies usually have major effects on the expression or function of the relevant gene. The consequences for immunity are correspondingly severe. In contrast to these rare, disruptive genetic abnormalities, there are minor, less impactful variations in gene sequence that occur more commonly in the human population and that can modify the immune response in subtle ways. Such genetic variants, which are referred to as polymorphisms rather than mutations, do not result in frank immunodeficiencies but they do have the potential to alter immune capabilities sufficiently to influence the outcome of infection. Polymorphisms in a number of genes involved in the immune response, including the genes encoding MHC molecules, have been implicated in susceptibility to infectious diseases like leprosy and tuberculosis, although their individual effects on disease risk are generally quite small.

Too Much of a Good Thing

On the flip side of immunodeficiencies are abnormalities that result from misdirected or excessive and unregulated immune responses (see figure 7.1, *right*). The autoimmune diseases, which represent one manifestation of the inappropriate targeting of immune functions, are a mixed bag of tissue-damaging disorders, their unifying feature being the development of an adaptive response against self antigens. Both B and T lymphocytes contribute to the autoimmune process, although their relative importance varies with the specific disease.

In some of the disorders, the pathologic process is largely driven by the production of autoantibodies that are directed against a single or small number of specific self antigens present in one tissue or cell type. An example of this type of abnormality is myasthenia gravis, a disease in which antibodies develop against a component of the neuromuscular junction, the structure that conveys signals between nerves and muscles. The binding of these autoantibodies to their target inhibits transmission of the signals and prevents normal contraction of muscles, particularly those controlling facial expression and eye movement. Drooping of the eyelids and double vision are characteristic clinical features. Another example is Graves' disease, in which antibodies against a receptor on cells of the thyroid activate the gland to produce excessive amounts of the hormone thyroxine, leading to metabolic disturbances, tremor, and protrusion of the eyes.

SLE, or lupus, is another autoimmune disease associated with the development of autoantibodies. In this condition, the antibodies are directed not against a single specific target but against a group of proteins that are usually found in the nucleus of most cells. Such proteins can be released into the extracellular environment following the death of cells, an event that occurs at a relatively low level under normal circumstances, and they can induce an antibody response in predisposed individuals. The binding of the antibodies to their target antigens leads to the formation of circulating antigen-antibody complexes (also known as immune complexes) that are ultimately deposited in the small blood vessels of various tissues, including in the skin, joints, and kidneys. As a result, the complement cascade is activated at these sites, initiating a chronic inflammatory process that damages the tissues and produces symptoms such as rashes, joint swelling and pain, and kidney failure.

In rheumatoid arthritis, B and T cells are activated by self proteins that become modified by the unusual amino acid citrulline, something that can happen during a transient inflammatory event. Since the modification does not occur under normal circumstances, B and T cells that carry antigen receptors with the ability to recognize citrullinated proteins are not eliminated by negative selection and may be activated if the conditions are right. Immune targeting of such modified self antigens, which may be particularly abundant in the lining of joints, leads to chronic inflammation of this tissue and, ultimately, to destruction of the adjoining bone. The inflammatory process manifests as pain, stiffness, and swelling of the affected joints. If it is uncontrolled, it can lead to disabling skeletal deformities.

Other autoimmune diseases appear to result mainly from the rogue activation of T cells, although autoantibodies may also be involved. Examples of this type of problem include multiple sclerosis and IBD. These conditions are characterized by T cell-mediated chronic inflammation in the central nervous system and gastrointestinal tract, respectively, and lead to symptoms specific to those tissues: episodic and recurrent neurologic deficits in multiple sclerosis, and abdominal pain and bloody diarrhea in IBD. In multiple sclerosis, the antigens that incite the abnormal immune activation are components of the fatty sheath (myelin) that surrounds the long extensions or axons of nerve cells. The target antigens in IBD are not well defined but are probably derived from the gut microbiota.

Genes and Autoimmunity

Why does the immune system suddenly turn traitor in some individuals and start to attack the body's own cells and molecules? Unfortunately, we do not yet have a good answer to that question. In very broad terms, autoimmune diseases result from a failure of the mechanisms that normally prevent self-reactive B and T lymphocytes from developing or being activated. As in the immunodeficiencies, a small fraction of the diseases is attributable to mutations in single genes that are involved in these mechanisms or that regulate other aspects of adaptive immunity and inflammation. One such gene is called AIRE and is required for negative selection of developing T cells in the thymus. Individuals who have two copies of a mutated and nonfunctional AIRE gene are unable to weed out self-reactive T cells from their mature repertoire. Despite the other checks and balances that are designed to prevent autoreactivity, these cells become activated by self antigens, most commonly in endocrine organs such as the adrenal and parathyroid glands, with consequent tissue damage and the development of corresponding clinical abnormalities. A similar autoimmune disease, involving multiple tissues, occurs in individuals with mutations in the FOXP3 gene, which is required for normal differentiation of regulatory T cells or Tregs. In the absence of the "braking" function of Tregs, various components of the immune system run amok, leading to widespread tissue inflammation.

Most autoimmune diseases, including relatively common ones like IBD and lupus, cannot be linked to a single gene defect. But many of them tend to run in families. If you have a parent or sibling with IBD, for instance, you are three to twenty times more likely to get the disease yourself than someone in the general population. Your risk is even greater if both of your parents or your identical twin are affected. These and other observations point to a clear genetic contribution to the development of such diseases. While the exact nature of this contribution remains to be clarified, the interplay of commonly occurring genetic polymorphisms is thought to play a role.

IBD has been particularly informative about the role played by gene variants. The disease occurs in two major clinical forms, ulcerative colitis and Crohn's disease, which affect different parts of the gastrointestinal tract and have distinct pathologies and manifestations. At latest count, more than two hundred genetic polymorphisms have been linked to ulcerative colitis, Crohn's disease, or to both forms of IBD. The functions of the affected genes provide some clues about how the disease might develop. Several of the variants are in genes involved in maintaining the barrier or antimicrobial properties of the intestinal epithelium. Impairment of these functions could lead to increased exposure of subepithelial immune cells to components of the microbiota and consequent activation of inflammatory responses. Other IBD-associated gene polymorphisms influence the activation or regulation of the antimicrobial and inflammatory functions of macrophages and other immune cells, and could lead to the cells responding inadequately, excessively, or inappropriately to microbial intruders. In keeping with this idea, some of the gene variants that have been linked to IBD have also been implicated in the development of infectious diseases such as leprosy. Based on the insights provided by the study of gene variants, and on supporting evidence from animal experiments, current thinking holds that IBD is the consequence of an abnormal, unregulated immune response to components of the microbiota. The disease appears to represent a breakdown of the truce that normally exists between microbiota and host, although much remains to be learned about how this occurs. In particular, connecting specific gene polymorphisms to alterations in immune function and from there to the pathologic and clinical abnormalities of IBD is the focus of intensive, ongoing research.

You Win Some, You Lose Some

There is a striking discordance between the sexes in the occurrence of autoimmune diseases: disorders like SLE, rheumatoid arthritis, and multiple sclerosis are three to seven times more common in females than in males, a difference that holds true in both humans and laboratory mice. The exact mechanisms that are involved in this difference are not well understood, although several have been implicated. Variations in the levels of sex hormones like estrogen are probably important since immune cells are known to be sensitive to hormonal influences. Another factor that has recently come to light is the role played by the X chromosome.

Men have only one X while women have two, a distinction that would be expected to result in the presence of twice as many X chromosome genes in female cells as in male cells. To prevent the potentially detrimental consequences of this difference in dosage of X-linked genes, one of the X chromosomes in every female cell is subjected to a process of inactivation in which almost all the genes on that chromosome are silenced. However, recent experiments have revealed that this X chromosome inactivation is not perfect. It fails to occur in a small, variable fraction of some cells of the immune system, including monocytes, dendritic cells, and B cells, resulting in the disinhibition of genes that should have been kept suppressed. As a consequence, the affected cells have two times more of these genes that are active than they should, leading to an increase in the levels of the corresponding mRNAs and proteins. Several of these proteins are involved in microbial sensing, B cell signaling, T cell recruitment, and the production of type 1 interferons. If X inactivation is compromised in a sufficient number of cells, the elevated expression of the disinhibited genes could make the female immune system more responsive when stimulated. The heightened responsiveness has been demonstrated in some experiments and has been proposed as one of the mechanisms that contributes to the higher rates of autoimmune disease in women.

There is a satisfying parsimony to this mechanism since the increased expression of X chromosome genes that play roles in immune activation also provides an explanation for the well-established observation that females are considerably more resistant to infection than males (the greater severity of COVID-19 in men than in women may be one manifestation of this difference). Corresponding to their enhanced resistance to pathogens, females generally have higher levels of circulating antibodies and immune cells, and they respond better to vaccines and other challenges. It appears that evolution has endowed women with an immune system that is more potent than in men, presumably because of the transgenerational survival benefits associated with the placental transfer of antibodies and the dominant role played by women in child-rearing. Unfortunately, this gift is like the Trojan horse since it comes with an elevated risk of autoimmunity.

Allergic Diseases: Immune-Environment Disharmony

Allergies represent the unpleasant manifestations of an inappropriate immune response to a normally harmless substance in the environment. The manifestations are protean in nature, ranging from the runny nose and watering eyes of a seasonal pollen allergy to the severe systemic reaction brought on by an allergy to bee stings. As is the case for IBD and some other immune-mediated diseases, the number of individuals who suffer from allergies has increased significantly in recent decades, particularly in high-income countries. Most parents of school-age children have probably experienced the effects of this trend in the form of heightened and widespread concerns about peanut allergy, as well as the peanut-free lunch tables and restrictions on sharing snacks that schools have implemented to mitigate these concerns.

One of the more common types of allergy is mediated by an abnormal IgE response. The environmental instigator of the response can be any one of a long list of things that come in contact with the skin, gastrointestinal tract, or airways and include foods such as peanuts or shellfish, pollen, animal dander, dust mites, and the venom of bees and wasps. Most people do not suffer any harm from exposure to these substances (apart from the pain and annoyance caused by an insect sting), but in a minority of individuals they can provoke reactions that can be merely inconvenient—hives and itching—or potentially life-threatening: narrowing of the airways and collapse of the circulation. These responses typically occur within minutes to hours of the exposure and are usually caused by preexisting antibodies of the IgE isotype that have developed against the environmental trigger in the affected individual.

As we discussed in an earlier chapter, IgE is usually produced against intestinal parasitic worms and certain types of venoms; in that setting, it may provide protection without significant untoward effects. But for reasons that are not clear, some individuals aberrantly generate high levels of IgE antibodies against innocuous environmental antigens, including some found in food. Once it is produced, the IgE becomes tightly bound to high-affinity IgE receptors on the surface of mast cells. When the IgE is exposed to its target antigen in the peanut or shellfish or bee venom (often referred to as an allergen in this context), it activates the mast cell to unload a large cache of bioactive molecules that are stored within cytoplasmic granules into the surrounding tissues. These molecules, which include histamine, prostaglandins, leukotrienes, protein-degrading enzymes, and TNFα, have several effects: increased blood flow and fluid leak into local tissues, contraction of muscles of the gut and airways, induction of mucus secretion by intestinal and respiratory epithelial cells, and stimulation of nerve endings.

The location and magnitude of mast cell degranulation determine the symptoms that result. If the mast cell activation is restricted to the skin, for instance, the allergy will manifest mainly as hives, a consequence of fluid leaking from blood vessels into tissue spaces, and itching, a result of irritation of certain types of nerves. If it occurs in the intestine, the same events, along with contractions of gut muscles, may lead to nausea, vomiting, abdominal cramps, or diarrhea. Contraction of airway muscles may be felt as difficulty breathing, while leaking of fluid into the tissues of the larynx can produce a dangerous narrowing of the upper airway that can impede the intake of air. Combinations of the various symptoms can occur if mast cell activation takes place in multiple tissues. In addition, a potentially life-threatening consequence of allergic reactions is anaphylactic shock, which is the result of widespread blood vessel dilation and leakiness, leading to a drop in blood pressure and compromised circulation. In such situations, prompt administration of epinephrine, either through a personal automated injection device or by some other means, can be lifesaving. Epinephrine reverses many of the harmful effects of the mast cell molecules by constricting and reducing the permeability of blood vessels, and by relaxing airway muscles. Antihistamines can be used to relieve minor allergic symptoms such as hives and itching. And, of course, it is a good idea to avoid exposure to the offending substance in the future.

IgE Is Not the Only Culprit

Not all allergic responses to environmental antigens are mediated by IgE. Cellular components of the immune system, eosinophils and T cells in particular, are important players in certain types of abnormal immune reactions to food, the former in a group of illnesses known as eosinophilic gastrointestinal disease (EGID) and the latter in celiac disease. EGID is characterized by the dense accumulation of eosinophils, a rare type of blood cell recognized by its typical pink-staining cytoplasmic granules, in specific regions of the gastrointestinal tract. The eosinophilic infiltration produces symptoms such as difficulty swallowing, diarrhea, and rectal bleeding that vary with the region affected. Milk, eggs, legumes, nuts, and seafoods are the usual foods that provoke the eosinophilic pathology. T cells are the main culprits in celiac disease. They orchestrate a chronic inflammation of the intestine following recognition of antigenic peptides derived from gluten, a protein found in cereals like wheat, rye, and barley. The inflammatory process leads to symptoms such as abdominal pain, diarrhea, and weight loss. In both EGID and celiac disease, the symptoms can be alleviated and prevented by avoiding the relevant foods.

Tolerance Is a Virtue

The development of allergies to foods is particularly puzzling when you consider that ingested material normally *suppresses* immune responses. This lack of reactivity to dietary substances, a phenomenon known as oral tolerance, is necessary for us to consume and benefit from food, which, after all, is made up of a large number of foreign antigens that may be recognized by our B and T lymphocytes. The fact that these cells do not respond to food is owing to the operation of several important mechanisms that together lead to the development of oral tolerance.

Most of the food that we eat is completely digested into component molecules—sugars, amino acids, fatty acids, and so on—that are intrinsically rather ineffective at eliciting an immune response. Small amounts of undigested or partially digested food molecules that could be seen as foreign antigens may be absorbed, including proteins and short peptides. However, they do not usually activate PRRs and thus do not induce the dendritic cells that present the antigen to express the co-stimulatory molecules needed for T cell activation. In fact, T cells that recognize the antigen in the absence of co-stimulation (Signal 2) will either undergo programmed cell death or will become unresponsive to further stimulation as a result of entering the state of anergy. In addition, food antigens can stimulate the differentiation of Tregs, the T cell subset that inhibits the functions of other cells of the immune system. The Tregs, together with the induction of T cell death and anergy, ensure that most of us can eat whatever we want, including all the peanuts and shellfish that we might have a craving for, without suffering any ill effects.

Genes and Allergy

If oral tolerance is the usual response to ingested substances, why do certain foods provoke an allergic reaction and cause discomfort, illness, or even death in some people? There are a very small number of allergic individuals, including some with food allergies, who have single-gene mutations that help to explain their problems. One such rare mutation is in the filaggrin gene, which is involved in maintaining the integrity of the skin epithelium. Mutation of the filaggrin gene compromises the barrier properties of the skin and increases the risk of developing eczema early in childhood. Eczema is an itchy, inflammatory condition that is associated with an abundance of Th2-promoting cytokines. So, if a peanut antigen, for instance, enters through the eczematous skin (which is possible since traces of food can be found free-floating in the environment), the local cytokine milieu will favor the development of a Th2-dominated immune response, including production of peanut-specific IgE antibodies, rather than the oral tolerance that would have been induced if the peanut had been ingested. The next time that individual eats something containing peanuts, the binding of peanut antigen to the preformed IgE will trigger mast cells to release their stores of histamine and other molecules, producing the allergic manifestations we have seen earlier. Similar mechanisms are probably involved in the increased susceptibility to allergies seen in children who have eczema that is not associated with filaggrin mutations.

The vast majority of people with allergies do not have single-gene defects. However, genetic variation does play a role in IgE-mediated allergic disorders. Several studies have identified a large number of gene polymorphisms that are associated with a heightened risk of developing such disorders. Many of them have the potential to influence various aspects of the immune and inflammatory responses, particularly in relation to the differentiation of Th2 cells and the production of IgE. Variants in genes involved in epithelial barrier properties have also been found to affect allergic risk and may act similarly to mutations of filaggrin. As in IBD, the effect of each gene variant is relatively small, but it is presumed to increase the likelihood that Th2-skewed immunity and IgE antibodies will develop against a normally innocuous antigen, setting the stage for an allergic response if the antigen is subsequently encountered. The role of genetics in allergic problems that are not related to IgE and in reactions to food that are associated with eosinophilic infiltration is uncertain. Celiac disease, however, has a well-established connection to genes encoding specific class II MHC molecules: almost all those affected express the DQ2 or DQ8 versions of the molecules, which are required to present gluten-derived antigenic peptides.

A Peanut a Day Keeps the Allergist Away

Our understanding of immunology has had a significant impact on the practical management of food allergies, particularly peanut allergy. Previously, the official guidance was to avoid feeding young children anything with peanuts so that they would not develop an IgE response to the nut antigens. Now that we know that antigens administered via the gastrointestinal tract induce a state of tolerance, current recommendations favor early exposure to peanut-containing foods as a way of preventing the sensitization and IgE production that might occur if the immune system encountered peanut antigens through a route other than the intestine. As an extension of this idea, one strategy for the treatment of established allergies involves ingesting the allergen, or placing it under the tongue in tiny, slowly increasing doses, the goal being to induce enough oral tolerance that accidental or deliberate exposure to a small amount of the offending substance does not provoke a potentially life-threatening anaphylactic reaction. The exact mechanism by which this desensitization is achieved is not clear but has been suggested to involve the induction of allergen-specific Tregs and antibody isotypes other than IgE.

A Broader View of Abnormalities of the Immune System

Before concluding our discussion of the diseases of immune function, it is relevant to point out that the immune system does not exist in isolation. It is closely intertwined with other aspects of physiology, including general metabolic processes and the functions of specific organ systems. In fact, it is becoming increasingly apparent that immune cells actively participate in the function of various tissues under basal conditions and that their contributions in some cases have nothing to do with antimicrobial defense. A good example of this type of interaction is the involvement of microglia—the brain version of macrophages—in synaptic pruning, a process in which the junctions between nerve cells are sculpted by a nibbling form of phagocytosis. The pruning modulates interneuronal connectivity and plays an important role in normal brain development and in cognitive functions like learning and memory.

The intimate links between the immune system and the rest of the body can be a mediator of pathology: disturbances of immune cell function can have adverse effects on other tissues and, conversely, abnormalities of those tissues can lead to damaging immune responses. A new awareness of this cross-talk and its potential for setting up a dangerous vicious cycle has changed our view of many disorders that traditionally have not been associated with the immune system, including obesity, atherosclerosis, diabetes, and Alzheimer's disease. Obesity, for instance, is now considered to involve a state of chronic, low-grade inflammation that may be initiated by the accumulation of lipids and their metabolic products in fat depots, the liver, and other tissues. The inflammation, in turn, has been implicated in some of the metabolic derangements that occur in obesity, such as insulin resistance, type 2 diabetes, and fatty liver. Chronic inflammation is also seen in atherosclerosis, diabetes, and Alzheimer's, but how exactly it develops and how it contributes to the disease process in these situations remain unresolved. Because of the bidirectionality of the cross-talk between the immune system and other physiologic and metabolic functions, it is often difficult to disentangle cause and effect and pinpoint the key initiating events in the pathologic process. Clarifying the mechanisms that are involved and finding ways to manipulate them beneficially are active areas of research. The results of these studies could help in the prevention and treatment of problems that affect millions of people worldwide.

"The Fault, Dear Brutus, Is Not [Only] in Our Stars"

Can the immune-mediated diseases that we develop and the form that they take be explained entirely by our DNA? Certainly, that is part of the story, and it would be satisfying to stop here by concluding that variations in genes involved in the immune response are the basis for the development of immunodeficiencies, autoimmune and allergic disorders, at least some infectious diseases, and perhaps even problems like atherosclerosis and diabetes. That conclusion would be incorrect, however, since genetic variation accounts for only a part, often a minor part, of the variation in the occurrence or form of these problems. For instance, most individuals who carry the DQ2 or DQ8 forms of the class II MHC molecule do not suffer from celiac disease even if they consume normal quantities of gluten-containing foods. These gene variants, along with gluten, are *necessary* but not *sufficient* for developing celiac disease. Moreover, the increases in allergies and autoimmune disorders that have been observed in recent decades, particularly in people living in high-income countries, cannot be explained by changes in the genome since such changes do not occur on that timescale. Additional, nongenetic effects are clearly at play.

In some cases, the nongenetic factor is an intercurrent infection, usually with a virus, that pushes an otherwise healthy individual who happens to carry a genetic risk variant for IBD or lupus or some other chronic autoimmune problem to develop overt disease. Recent work has implicated infection with Epstein-Barr virus, the pathogen that causes infectious mononucleosis or "mono," as a key event in the development of multiple sclerosis. The data indicate that in some genetically predisposed individuals, proteins expressed by the virus induce antibodies and T cells that cross-react with self proteins, including a component of myelin and an adhesion molecule expressed by accessory cells of the nervous system. Interactions between the cross-reactive antibodies and T lymphocytes and their target antigens are presumed to initiate an inflammatory process that contributes to disease development.

Epidemiological studies have also shown that the incidence and course of many immune-mediated disorders are influenced by things that have nothing to do with DNA, including mode of birth, infant feeding practices, location of residence, diet, exercise, smoking, consumption of alcohol, and even the presence of a pet in the household. These studies suggest, somewhat surprisingly, that environmental and lifestyle factors can affect your immune system and the likelihood that you will develop IBD, asthma, or a food allergy as much as, perhaps even more than, the genes that you inherit. That conclusion immediately raises an interesting question: how can your environment and lifestyle influence the function of your immune system and your risk of developing an immune-mediated disease? Although the question does not have a definitive answer currently, some ideas worth considering may emerge as we continue our discussion.

• CHAPTER 8

Conditioning of the Immune System by the Microbiota

About three hundred and fifty years ago, Antonie van Leeuwenhoek, a Dutch lens maker and amateur scientist, used the microscope that he had painstakingly crafted to examine samples of scrapings from his mouth and a bit of his own feces. Where you and I might have recoiled in disgust, he was delighted and amazed to see many tiny moving creatures, which he called animalcules and which we now know were bacteria. His description of the findings in a letter to the Royal Society of London probably represents the first documentation of the existence of the microbiota, the community of microorganisms—bacteria, viruses, and fungi—that live on the skin or mucosal surfaces. Over the course of the many years since van Leeuwenhoek's observations, these organisms continued to attract scientific interest. However, their contributions to the physiology of their hosts, including the functioning of the immune system, were not widely appreciated until around the early 2000s.

Getting to Know Our Resident Microbes

For many years, studies of the microbiota were conducted largely outside the limelight. Nevertheless, important advances were being made, particularly the development of methods to characterize the constituents of resident microbial communities. An early milestone was crossed in the 1940s with the successful growth of intestinal bacteria in the laboratory using specialized culture techniques. Most of these organisms cannot survive in the presence of oxygen, so methods had to be devised to avoid exposure to this abundant atmospheric gas. Even with the use of oxygen-depleted conditions, the majority of the microbiota remained unculturable until recently, requiring the application of DNA sequence-based approaches to obtain a complete picture of the diversity of the tissue-resident microbial communities. These techniques do not depend on actually culturing organisms and rely instead on sequencing DNA isolated from biological samples to identify the associated microorganisms. Another key step forward was the generation of mice and rats that completely lacked a microbiota. Such germ-free animals are derived by delivering the rodent pups by cesarean section under sterile conditions, and then rearing them in special isolators that are completely devoid of microorganisms. Experiments with germ-free mice demonstrated unambiguously that the microbiota had a significant impact on various aspects of host physiology, including processing of nutrients and medications, bile salt metabolism, weight gain, and the accumulation of fat.

Not Just Neighbors

Interesting as these developments were, they probably went unnoticed by most scientists, including immunologists. However, they provided the essential foundation for two landmark studies that drew the attention of a large section of the biomedical research community. One set of experiments, published in 2004, was spearheaded by Seth Rakoff-Nahoum, who was then a graduate student in the laboratory of Ruslan Medzhitov, a well-known immunologist at Yale University. Rakoff-Nahoum's observations in a mouse model of colonic inflammation showed that the gut microbiota continuously activated host TLRs, inducing signals that promoted the repair of the intestinal epithelium following injury. In the absence of these signals, the repair process was impaired and the epithelial injury resulted in significant pathology and even death. The results of the experiments were dramatic. Mice that were able to interact with their microbiota through TLRs recovered quite quickly from chemically induced damage to the epithelial lining of the colon. In contrast, mice that had been depleted of their microbiota by antibiotic treatment or that had been genetically engineered to lack the TLRs needed to perceive microbiota-derived signals failed to heal the damage and died following treatment with the chemical. The results pointed to the inescapable conclusion that the intestinal microbiota communicated with host cells through TLRs and that this interaction was required for epithelial repair. These experiments indicated for the first time that the gut-resident microbial community was an essential component of a host homeostatic mechanism, one that helped to maintain and restore the intestinal epithelial barrier, a key element of frontline antimicrobial defense. In a sense, the microbiota acted in this situation as an intrinsic part of the host, almost like another organ or tissue.

The notion that host and microbiota were functionally interconnected was taken a step further by another PhD student, Peter Turnbaugh, who was working in the laboratory of Jeffrey Gordon at Washington University, in St. Louis. In studies reported in 2006, Turnbaugh found that there were clear differences in the composition of the intestinal microbiota between lean and obese mice, and between lean and obese humans. Moreover, germ-free mice that were colonized with gut microbiota from obese animals accumulated a greater amount of body fat than mice colonized with the microbiota of lean animals, even though both of the colonized groups ate similar amounts of the same diet. These striking findings indicated that the microbiota was sufficient to confer an important health-related characteristic, the ability to harvest energy from the diet and convert it into fat. The microbiota was thus a transmissible determinant of host phenotype. This was a remarkable conclusion, particularly at the time the paper was published, tantamount to saying that the color of your eyes depended on the kind of bacteria that lived in your gut.

The Microbiota Can Do Anything and Everything

The Rakoff-Nahoum and Turnbaugh papers marked the start of a revolution in microbiota research, a hugely productive period that has yielded a treasure trove of data and a substantial increase in understanding of the microbiota and its effects on the host. To get some idea of how explosively the field has grown, you only have to search the PubMed database of scientific literature with the term "microbiota." The results will show that a mere 1,400 papers related to our resident microbes were published in the forty-nine-year period between 1956 and 2005, whereas there were close to 81,000 papers on the subject in the sixteen years from 2005 to 2021! Many of the investigations that the more recent of these publications represent used the experimental strategies pioneered by the Gordon lab: first comparing the composition of fecal or other microbial communities between groups of individuals that differed in some interesting way, and then using colonization of germ-free mice to test the role of the microbiota in the differences. Nowadays, if the involvement of the microbiota is confirmed by such studies, the initial experiments are often followed by genetic, biochemical, and other analyses to identify the specific microbes and microbial molecules responsible for the difference in phenotype.

The results of innumerable studies designed broadly along those lines have revealed links between the microbiota and almost every aspect of mammalian physiology in health and disease, including nutritional status, lipid and carbohydrate metabolism, the function of multiple tissues, the disposition and efficacy of various drugs, diurnal rhythms, cognition, behavior, mood, and the response to antitumor therapy. The list of effects attributed to the members of our resident microbial communities is now so long that one of my more cynical colleagues was heard to remark irritably one day that about the only thing that the microbiota has *not* been credited with is doing the laundry.

Microbiota-Immune System Communication: PAMPs, Microbial Antigens, Metabolites

Nowhere is the influence of the microbiota more apparent than in the immune system (figure 8.1). Studies comparing normal mice and mice that either lack a microbiota or contain an altered microbiota

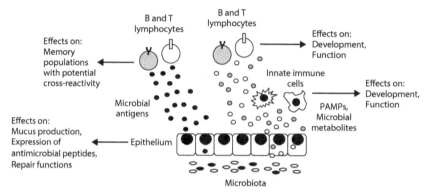

FIGURE 8.1. Effects of microbiota-derived molecules (PAMPs, metabolites, antigens) on innate and adaptive components of the immune system.

have highlighted the numerous ways in which the microorganisms that reside on the skin or mucosal surfaces influence immune cell populations and functions. We have already learned from the work of Rakoff-Nahoum and his colleagues that the resident microbiota contributes to the integrity of the first line of immune defense in the gut, the intestinal epithelial barrier. Subsequent studies by other investigators have confirmed this observation and have extended it to other aspects of barrier function such as the production of mucus and antimicrobial peptides. Similar effects of resident bacteria on the maintenance and repair of the epithelium have been documented in the skin, and they probably also occur in the respiratory and genitourinary systems.

Beyond the epithelium, the microbiota has been shown to affect the development or function of many other cells of the immune system, including macrophages, dendritic cells, neutrophils, innate lymphocytes, B cells, and T cells. These effects are particularly prominent in tissues such as the intestine and skin, which are in close proximity to resident microbial communities. But they also influence immune cells in more distal sites like the lymph nodes, the circulation, and the bone marrow. In some cases, specific members of the microbiota have been linked to the number or properties of different immune cell populations. For example, a group of distinctively shaped microbes known as segmented filamentous bacteria promote the development of the Th17 subset of helper T cells in the small intestine of mice, while several types of colon-resident bacteria induce regulatory T cells (Tregs).

A variety of mechanisms have been shown to mediate the effects of the microbiota on the immune system. Most of them involve molecules that are released by the resident microbes and that then diffuse or are transported across the epithelium of the skin or mucosal surfaces to underlying and distant tissues. These molecules include PAMPs, microbial antigens, and various metabolites, all of which can alter the behavior of host cells that they come in contact with. For instance, LPS, a PAMP produced by Gram-negative bacteria in the gut, can get across the intestinal epithelium in small quantities and influence the properties of macrophages, dendritic cells, neutrophils, and other immune cells by chronically stimulating TLR4 on these cells. Similarly, antigens derived from the microbiota can activate B and T lymphocytes through their antigen receptors, leading to the production of antibodies and effector T cells and, ultimately, to a population of memory lymphocytes with reactivity to organisms that express those antigens.

One of the recent developments in the understanding of microbiota–immune system interactions relates to the role played by microbial metabolites. Bacterial members of the microbiota (and probably fungal components too) are able to metabolize molecules derived from the diet or host cells, including carbohydrates, amino acids, and lipids, to produce a variety of relatively small organic compounds that can have profound effects on immune cells. The short-chain fatty acids (SCFAs) acetate, butyrate, and propionate represent a particularly interesting class of these metabolites. They are produced by the breakdown of dietary fiber by certain types of gut-resident bacteria and have been shown to promote the differentiation of Tregs and memory T cells, alter ILC function, and enhance the antimicrobial capacity of macrophages.

The influence of SCFAs on the immune system is one explanation for the benefits of a high-fiber diet. SCFAs bring about their effects by interacting with specific receptors on host cells or by modifying chromatin proteins associated with the genomic DNA of the cells. Besides SCFAs, breakdown products of the amino acid tryptophan represent another group of bacterial metabolites that have been shown to influence the functions of various immune cells.

As alluded to in earlier chapters, many of the effects of the microbiota on the immune system are involved in maintaining peaceful coexistence with the host. Microbiota-induced responses—such as the expression of mucus, antimicrobial peptides, and other barrier-reinforcing molecules by epithelial cells; the production of microbiota-reactive IgA by mucosal B cells; and the generation of Tregs against antigens of the resident microbiota—can all be viewed in this light. Each of these responses serves to minimize contact between resident microorganisms and host cells, or to dampen the inflammation that might result from such contact. Although they may have evolved to preserve the host-microbiota détente, they can also influence susceptibility to infections and other diseases associated with the immune system. A deficient or abnormal microbiota can affect the expression of epithelial antimicrobial peptides, for instance, or the number of Tregs and Th17 cells, and those effects in turn can influence the response to a pathogen or other challenge to the immune system. Levels of gut antimicrobial peptides that are too low may increase the likelihood that you will get sick from a *Salmonella* infection, while elevated numbers of Th17 cells or too few Tregs may put you at risk of developing a chronic inflammatory disorder like IBD. This concept—that the microbiota is an important factor in the development and course of infectious and immune-mediated diseases because of its effects on immune function—has been substantiated by a large body of work in mice and is also supported by correlative data from humans. A few examples that illustrate the experimental basis for the concept are provided in the next section.

Tales from the Toilet

Many studies of the microbiota are based on characterizing the microbial communities in stool samples, most often by using DNA sequence information to identify the microbes present. A recent analysis of this type found that there were significant variations in fecal microbiota composition between groups of healthy people from the United States, Fiji, and Guatemala. To assess the functional consequences of this difference, the study investigators colonized groups of germ-free mice with fecal material obtained from donors representative of the three human populations, and then challenged the mice with an intestinal pathogen similar to the organism that causes traveler's diarrhea. The results of the experiment showed clear differences in susceptibility to the pathogen, with the "Guatemalan" mice being most resistant, the "Fijian" ones most susceptible and the "US" ones showing an intermediate phenotype. The resistance of the mice colonized with the fecal material from Guatemalan donors was associated with elevations in the intestinal levels of two antimicrobial peptides, as well as the cytokine interferon γ, providing a potential explanation of the observations.

Another experiment used similar approaches to examine the relationship between the gut microbiota and the development of asthma, an allergic respiratory disease characterized by episodic cough and breathing difficulty resulting from inflammation and constriction of the airways. It was based on a cohort of healthy children who were monitored from birth to school entry, including periodic clinical evaluations and collection of fecal samples. Some of the children developed allergic manifestations and asthma by the age of five years. When the investigators leading the study characterized the stool microbiota, they made a striking observation: the children who developed asthma had significantly lower levels of four types of bacteria compared to the children who did not develop asthma, a difference that was noted even in the fecal samples that had been collected at the age of three months, well before the onset of any allergic or asthmatic symptoms. The investigators colonized germ-free mice with fecal material representative of either the asthmatic or non-asthmatic children and tested the animals in a model of respiratory allergy that resembled asthma. Sure enough, the mice that received the feces from the asthmatic children developed more severe lung pathology than the ones that got the stool from the non-asthmatic children. Although the mechanism of this effect was not explored in detail, the investigators did note that the stool of the children who developed asthma had significantly lower levels of acetate, one of the SCFAs known to have potent immunomodulatory effects.

In a final illustration of how the microbiota can influence the immune response, several research groups have shown that most healthy individuals have circulating memory lymphocytes directed against antigens expressed by the resident microbiota of the gut. Presumably, these cells were generated when the microbiota antigens leaked across a temporary breach in the intestinal epithelium, which could have developed, for instance, during an episode of gastroenteritis. Surprisingly, some of these memory B and T lymphocytes cross-reacted with pathogens that had never been encountered previously by the individuals who had these cells in their blood, including the AIDS virus and the malaria parasite. Not only that, the microbiota-induced memory B lymphocytes were shown to affect the outcome of subsequent exposure to the cross-reactive antigen, decreasing the likelihood of being infected by the malaria parasite or skewing the response to an experimental HIV vaccine in a direction that reduced efficacy. The results suggested that having the microbiota composition that generated these cross-reactive memory lymphocyte populations could be helpful if you happened to live in a malaria-endemic region of the world, but it could also make it less likely that you would benefit from the HIV vaccine.

Lesser Known But No Less Important

The bulk of the microbiota at most tissue sites is made up of bacteria, so it is not surprising that these are the organisms that have been studied most intensively. The viral and fungal components of the microbiota, the virome and mycobiome, respectively, have so far received less attention than the bacterial members of the community. That situation is changing rapidly, however, and we are learning more and more about these members of our resident flora. Most of the intestinal virome, a community of close to a billion organisms per gram of stool, is made up of bacteriophages, viruses that infect bacteria. Any effects that they might have on immune function are probably related to their ability to influence the viability or metabolism of their hosts, the bacterial members of the microbiota. About 1% of gut-resident viruses infect mammalian cells, latently or productively. Recent experiments indicate that these eukaryotic viruses can have effects on the immune system similar to those of bacterial members of the microbiota and can influence susceptibility to infection and immune-mediated diseases in animal models.

Like the virome, the mycobiome is also attracting increasing interest. Fungi resident in the gastrointestinal tract and upper airway have been characterized using culture-dependent and culture-independent methods, and they have been linked to changes in immune function both in health and in disease states. In one recent study, specific strains of the resident fungus *Candida albicans* were implicated in inducing the production of the cytokine IL-1β and in promoting intestinal inflammation in both mouse models and in human ulcerative colitis. Future experiments will undoubtedly extend and clarify the immunological effects of the virome and mycobiome, so stay tuned!

Nurturing the Microbiota

The results of a large number of studies in mice and humans, including many like the ones that I have described, indicate clearly that variations in the microbiota matter. How do these variations arise? We all start as blank slates as far as our microbiota is concerned—the intrauterine environment is sterile for all practical purposes, and the fetus is generally considered to lack a resident microbial community (some recent analyses suggest that this idea may not be entirely correct, but the findings are currently quite controversial). The first microorganisms to colonize the newborn are those that are encountered during birth: the microbes associated with either the mother's birth canal or skin depending on whether the mode of delivery is vaginal or cesarean. The initial microbial colonists set the stage for later arrivals from the baby's environment, which generally take up residence in waves over the course of one to two years. By the time the child is about three years old, a core microbiota that resembles that of the adult in essential characteristics has been established on the skin and mucosal surfaces.

Aside from the core features, however, there is a great deal of variation in the composition and functionality of the resident microbiota. A number of factors are responsible for the differences. Mode of birth is one. The organisms that make up the maternal vaginal and skin microbial communities are not the same, and initial colonization by one or the other can influence the infant's subsequent microbiota composition (although the effects are generally most prominent shortly after birth and become more variable with age). Diet, starting with whether the infant is fed with breast milk or formula, is another determinant of microbiota composition. Human breast milk contains a number of unique complex carbohydrates, known as human milk oligosaccharides, that are not found in other types of milk or in formula. These carbohydrates act as nutrients that promote the growth of certain types of bacteria, particularly bifidobacteria, that have a number of health benefits, including on the immune system. Beyond infancy, diet continues to be a significant influence on microbiota composition, producing clear differences depending on whether an individual is a strict vegetarian, an ardent meat-lover, or something in between. Intercurrent illnesses, especially those that are associated with inflammatory changes in the skin or gastrointestinal tract, can produce marked alterations of the microbiota in those tissues. The use of certain medications, particularly antibiotics, can also cause dramatic shifts in microbiota composition. Other factors that have been shown to affect the microbiota, albeit in relatively minor ways, include exercise, smoking, alcohol consumption, urban or rural residence, and contact with household pets.

If that list of influences sounds familiar, it is because we have already come across it in the context of the nongenetic, environmental, and lifestyle factors that affect the occurrence and course of immune-mediated disorders. The overlap is not coincidental. In fact, it represents a potential solution to the puzzle that we were left with at the end of the last chapter: how to explain the effects of the nongenetic factors on such disorders. What the data suggest is that many of these effects may be mediated by alterations in the microbiota. Putting together what we learned in the last chapter with the observations discussed in the present one, the likelihood of an individual developing a food allergy, for instance, may be influenced by both the complement of genes that he or she inherits, including any genetic variants that affect allergy risk, and by environmental and lifestyle factors that adversely alter the microbiota. Both genetic and nongenetic influences, the latter potentially acting via the microbiota, may exert their effects by modulating immune function in a way that the probability of an allergic response is increased: perhaps by increasing the number of Th2 cells a little bit, or by decreasing the number of Tregs a tad, or by causing any number of subtle changes that make the response to a food antigen drift away from normal.

A Note of Caution

Despite the wealth of data linking changes in the microbiota to various disease states, it is important to mention that many of the findings in humans are based on correlations. Inferring causality from correlation is always fraught with risk. As an example, several studies carried out over the last five to ten years have demonstrated alterations in the fecal microbiota of individuals with autism. Even more interestingly, colonizing germ-free mice with stool from autistic donors induced changes in social behavior in the recipient animals, suggesting that the altered microbiota might be a cause of behaviors associated with autism in humans. As you might imagine, these results generated a lot of excitement. But a recent large-scale analysis of autistic and neurotypical children pointed to an alternate interpretation of the correlation between autism and microbiota alterations. The study presented evidence indicating that individuals with autism tend to have a less diverse diet than neurotypical subjects, presumably because of restricted eating preferences, and that the less diverse diet could explain the changes in the gut microbiota. So, although an altered microbiota can lead to autism-like behaviors in mice, the available evidence does not yet support drawing the same conclusion about humans.

But this outcome should not be used as a reason for repudiating or discouraging such studies. Rather, it just emphasizes the need to be open to the possibility that there might be different ways to explain experimental observations, especially when extrapolating from mice to humans. Moreover, the fact that the microbiota from autistic individuals influenced behavior when transplanted into germ-free mice remains undeniably interesting. If the mechanism of this effect can be worked out, it could provide insights into the links between diet, the microbiota, and the brain, and that information in turn could be used as the starting point for developing strategies to take advantage of those links for clinical benefit.

You Really Are What You Eat

The fact that the microbiota has the potential for profound effects on the immune system and on immune-mediated diseases can be viewed, and has been eagerly taken up, as a therapeutic opportunity. After all, it is easier to manipulate the composition of your resident microbial community than it is to modify your genes (currently, at least). This idea has been applied in a number of strategies to alter or make use of the microbiota for the purpose of preventing or treating various disease states.

One of the most useful approaches is fecal microbial transplantation, or FMT, which alters a disease-associated microbiota so that it more closely resembles the non-diseased state. The procedure involves the introduction of a slurry of feces from a healthy donor into the patient, usually through a nasogastric tube or a colonoscope. FMT has enjoyed spectacular success in the chronic, debilitating, and very hard to treat diarrhea caused by infection with *Clostridium (Clostridioides) difficile*, which typically occurs when the intestinal microbiota is depleted by prolonged antibiotic therapy. The exact mechanism of action of FMT is currently not clear, but it is at least partly attributable to inhibition of *C. difficile* colonization or growth. Although there is an unavoidable yuck factor associated with current versions of FMT, those who have benefited from it generally consider that a small price to pay for the almost miraculous disappearance of their symptoms. Moreover, recent refinements of the procedure, involving the administration of specific groups of beneficial microbes in capsule form, may soon make the use of fecal slurries obsolete. FMT is also being evaluated as a treatment for IBD. The results of some initial studies have been encouraging, but further work will be required to confirm and substantiate the findings.

Another widely applied strategy makes use of diets or dietary supplements that favor a normal microbiota or that correct an abnormal, dysbiotic one. The strategy can take various forms, from a general modification of the kinds of foods that are eaten—reducing the consumption of processed foods and increasing the intake of fruits and vegetables, for example—to the use of specific supplements that promote a healthy microbiota, including prebiotics (dietary components like fiber that encourage the growth of beneficial microorganisms), probiotics (yogurt and other fermented foods that contain live microorganisms), and postbiotics (microbial metabolic products like SCFAs that have anti-inflammatory and immunomodulatory effects).

The recommendation to increase consumption of fiber-rich foods such as whole grains, fruits, and vegetables is backed by solid scientific evidence. There is a large body of data from studies in human populations indicating that dietary fiber has a number of health benefits, including a lowered risk of developing cardiovascular disease and colon cancer, improvement of metabolic parameters such as insulin sensitivity, alleviation of the low-grade inflammation associated with obesity, and prevention of constipation. Many of the benefits are likely to be mediated by changes in the gut microbiota and, in some cases, by the consequent alterations of immune function. Experiments in laboratory mice have substantiated these ideas and have suggested immunological mechanisms that might be involved.

Another area where there appears to be growing agreement is regarding the need to reduce the intake of ultra-processed foods. The negative consequences of consuming ultra-processed foods have been highlighted by a recent study that examined over a hundred thousand adults from twenty-one different countries. The results of the analysis demonstrated a correlation between the risk of developing IBD and a diet rich in ultra-processed foods that contain emulsifiers, preservatives, artificial flavorings, or other chemical ingredients. Although the study did not analyze the impact of the diet on the microbiota, earlier work in mouse models of IBD has demonstrated that emulsifiers such as carboxymethylcellulose and polysorbate 80 can alter the composition of the gut microbiota in a manner that increases susceptibility to intestinal inflammation.

The evidence from studies analyzing the effects of diet on health certainly justifies the conclusion that consuming a fiber-rich diet is good for you and your microbiota, whereas foods containing a lot of chemical additives are potentially harmful. Unfortunately, it is difficult to be as sure about some of the other dietary interventions that are being promoted, including the use of probiotics, postbiotics, and prebiotics other than fiber. Some of these interventions have been shown to reduce the severity of the pathology in experimental models of infectious, inflammatory, and allergic disorders, but their efficacy in human disease is currently far from certain. The available data are conflicting or simply not compelling enough.

Not Everything Is about the Microbiota

Not all immunological effects of the diet and other lifestyle factors such as smoking and exercise are mediated by changes in resident microbial communities. Microbiota-independent effects on components of the immune system are also possible. For instance, dietary constituents like vitamins A and D and the minerals iron, zinc, and selenium can directly influence the function of various cells of the immune system in addition to their ability to alter microbiota composition. Inadequate intake of these micronutrients generally impairs immune responses, while dietary supplementation can improve them, especially if there is a preexisting deficiency. That some of these effects are directly on immune cells has been shown using genetically manipulated mice that lack molecules involved in the normal uptake, handling, or response to specific micronutrients. Rare mutations that affect micronutrient metabolism in humans have also been shown to alter immune function. For example, a mutation in the transferrin receptor gene that has been identified in a very small number of individuals prevents normal cellular uptake of iron and leads to a severe combined immunodeficiency affecting both B and T cells, highlighting the importance of this micronutrient in lymphocyte function.

The realization that certain dietary components might influence the occurrence or course of disease comes with a bit of irony (and comeuppance) for me. As a resident in pediatrics in India, I was often asked by the mothers or grandmothers of my patients whether a special item of food—usually a particular type of banana or tomato that they had been told had health benefits—might hasten the recovery of their child. Not wishing to encourage what I thought were their foolish beliefs, I am sorry to say that I always pooh-poohed their anxious questions and insisted that all they had to do was follow the course of "modern" treatment that I had prescribed. Although it is highly unlikely that these women had any knowledge of the microbiota or the immune system, I now have to concede that their faith in the healing power of food may not have been entirely misplaced, and that they could be well justified in saying, "I told you so."

CHAPTER 9

Remembrances of Things Past Haunt the Present

It is a truism that we are not who we used to be. From the moment of our birth, we are molded for better or worse by the people around us and by what we experience. Each encounter with our social environment leaves its mark, and each such mark influences the way we respond to subsequent encounters and experiences. So it is with the immune system.

The Importance of Being Wild

For much of its history, the field of immunology has relied on laboratory mice as the main experimental model. They are available in genetically pure strains, they can be bred easily, they can be manipulated and analyzed using a variety of technologies and reagents that have been developed specifically for the purpose, and they can be used to study many of the diseases that afflict humans. And, although you might not think it to look at them, they are one of our closest evolutionary relatives next to nonhuman primates. Certainly, experiments in mice have yielded important and broadly applicable insights into the functioning of the immune system and have provided a convenient preclinical testing ground for practical applications destined for human use. But there has been increasing concern that the mouse does not always perfectly replicate human physiology, especially after several treatment strategies that were successful in rodents failed to achieve the desired effects in humans.

With this concern in mind, David Masopust and his colleagues at the University of Minnesota carried out experiments a few years ago to find out how closely the mouse immune system mimics the human one. Previous studies had documented the overall similarity of the two systems with respect to organization, structure, cell populations, and functions. But the researchers noted some striking differences. Specifically, they found that the blood and tissues of adult humans were chock-full of memory T cells, presumably reflecting the outcome of multiple encounters with microbial and other antigenic stimuli, whereas adult mice were relatively devoid of these cells, resembling more closely in this respect the condition of newborn human babies. The observations suggested that the immune systems of adult laboratory mice lacked microbial and antigenic experience, a conclusion that was not too surprising in hindsight considering that the animals are almost always housed under ultraclean and carefully controlled conditions. Building on that idea, Masopust and his colleagues wondered whether the immunological characteristics of mice from the real world would be different.

To answer that question, the investigators had to descend from their ivory towers and go boldly into the wild. Well, not really into the wild, just to the nearest pet store, where they were able to procure a few mice with their credit cards, and then to a local farm, where they managed to catch some feral ones. (The details of the expedition were left out of the paper describing the experiments, but one imagines some hapless graduate student in a lab coat running around a hay-filled barn with a mousetrap while the farmer stands in a corner bemusedly shaking his head.) Once the pet store and feral mice were brought to the lab, a quick analysis revealed that their immune system parameters looked very different from their lab-raised cousins and a lot more like adult humans: many more circulating and tissue memory T cells and, as a later study showed, more phagocytes and inflammatory cytokines in their blood.

So, the researchers' hunch appeared to be correct: the immune system changed as it was stimulated by the microorganisms in the external environment, leading to alterations in the phenotypic characteristics and proportions of its cell populations. In keeping with this notion, the pet store and feral mice were teeming with bacterial, viral, and parasitic pathogens that were completely absent from the lab mice. Moreover, if the lab mice were co-housed in the same cage as pet store mice, a procedure that leads to sharing of resident microbes between cage-mates, their immune cells became progressively more mature in their characteristics and less like a newborn human's. It was possible that the microbiota was at least partially responsible for this effect since there were clear differences in the composition of the resident gut microbial communities between pet store and lab mice. However, a similar maturation of the immune system was seen if the lab mice were sequentially infected with three different chronic viruses and a parasitic worm, indicating that exposure to pathogens played an important role.

What Does Not Kill Us Makes Us Stronger

Clearly, it is not possible to do experiments like the ones in lab and pet store mice in humans: most institutional ethics boards would frown on deliberately infecting newborn babies with multiple viruses and parasites. But the higher numbers of memory T cells in adults relative to newborns suggests that the human immune system is also modified by infections and other forms of microbial contact. This idea is supported by the findings of a large-scale analysis of pairs of identical and nonidentical twins by Mark Davis's group at Stanford University. The study characterized interindividual differences in the blood levels of about two hundred different indicators of immunological status, including the frequencies of close to a hundred different cell populations and the concentrations of about fifty serum cytokines. The results showed greater and greater variability in these parameters between individuals as the subjects got older. This increase in variation was observed even when comparing one member of a pair of identical twins with the other, indicating that it was largely attributable to the effects of factors that were not inherited. One of the non-heritable influences that appeared to play a significant role in the variability of immune parameters was latent infection with cytomegalovirus (CMV), a virus that infects about 60% of the US population by the time of young adulthood and remains in a quiescent state within bone marrow precursor cells, usually without causing overt problems. If one member of a pair of identical twins was infected with CMV, there was less correlation between their immune parameters than if both members of the pair were uninfected, suggesting that exposure to the virus altered immune status.

A similar analysis was carried out recently to compare multiple immune system parameters in American and Bangladeshi children from birth to three years of age. The American kids were from the San Francisco Bay area while the Bangladeshi ones were from an impoverished region of Dhaka. As you would expect, the environment of the children in the latter cohort was such that they were exposed from an early age, especially after weaning, to multiple microbial pathogens. So, unlike the children in San Francisco, most of the individuals in Dhaka had suffered repeated bouts of diarrhea caused by intestinal viruses or parasites, and most had also been infected by CMV. Based on the parameters measured, the immune systems of the two cohorts showed clear differences. Overall, the characteristics of the T cells in the Bangladeshi children tended to resemble those of US adults, suggesting increased activation and accelerated maturation of immune status. While the study could not conclusively identify what was responsible for the variation in immune parameters between the San Francisco and Dhaka cohorts, differences in microbial exposure between the two locations was proposed as a potential explanation.

Both the mouse and human studies suggest that the immune system, as reflected by the characteristics and frequencies of its cell types and by the levels of various cytokines, undergoes broad changes in response to contact with infectious agents and other microorganisms in the environment. An important question raised by these observations is whether the interindividual variations in immune status that result from differences in microbial exposure have functional consequences. Do the variations actually matter in terms of the ability to mount an effective immune response? Experiments in mice indicate that they do, with data from human studies suggesting a similar conclusion. When groups of pet store and lab-raised mice were infected with *Listeria monocytogenes*, a pathogen that causes a flu-like illness with neurological complications, the former group of animals controlled the growth of the bacteria much better, ending up with ten thousand-fold lower numbers of *Listeria* in the liver than the latter. Similar results were obtained when the mice were challenged with a protozoan pathogen related to the malaria parasite. Since pet store and lab mice were of different strains and so had different genetic backgrounds, the experiments were repeated with groups of lab mice that had been co-housed or not co-housed with pet store mice for a few weeks. The outcome was the same: the co-housed mice controlled the *Listeria* infection better than the ones that had not been co-housed, suggesting that any differences in the genes carried by the pet store and lab mice were not a major contributor to the results. These findings indicated that being exposed to a microbe-rich environment, either directly or through the intermediary of a "dirty" animal, resulted in increased resistance to two very different types of pathogens.

Follow-up studies confirmed these observations, but they also raised the possibility that prior exposure to a heavy microbial load could be a disadvantage in some circumstances. In a model of sepsis in which the clinical abnormalities were largely determined by the severity of the systemic inflammation rather than by a specific infectious agent, lab mice that had been co-housed with pet store mice responded with higher circulating levels of several inflammatory cytokines and a correspondingly greater level of mortality than mice that had not been co-housed. In the context of a disease caused largely by inflammatory damage, it seemed that the revving up of the immune system that resulted from prior microbial exposure could be more harmful than helpful.

Blame It on Your Mother

Recent experiments carried out by Yasmine Belkaid's group at the National Institutes of Health indicate that pathogen exposure can influence immune function even before birth. In these studies, pregnant mice were infected about halfway through gestation with a weakened strain of the intestinal bacterial pathogen *Yersinia pseudotuberculosis*. The animals cleared the infection within a few days without suffering any ill effects. The *Yersinia* never crossed the placenta into the fetus, and healthy offspring were delivered at the end of the pregnancy. However, when the progeny of infected and uninfected mothers were compared at the age of 5–8 weeks (young adulthood for a mouse), clear differences in the immune system were discernible. Specifically, the mice born to *Yersinia*-infected mothers had significantly higher numbers of Th17 cells in the small intestine, indicating that the maternal exposure to the pathogen during pregnancy led to a persistent alteration in the immune system of the offspring. Not only that, this alteration had functional consequences: the increase in intestinal Th17 cells in the mice born to the infected mothers conferred enhanced resistance to infection with *Salmonella*, but it also made the animals more susceptible to an IBD-like intestinal inflammation.

These striking results clearly indicated that prenatal exposure to an infected environment led to a long-lasting functional alteration of the developing immune system of the fetus. So, it appeared that immune responsiveness could be influenced not only by an individual's own immunological history but also by his or her mother's experiences of microbial contact. What was a little puzzling, however, was how the maternal infection was influencing progeny immune function if the pathogen never crossed the placenta. Belkaid and her colleagues provided an answer to this question by showing that the pregnant mice responded to *Yersinia* infection with an increase in expression of the cytokine IL-6, and that the IL-6 made its way through the circulation into the fetus. The IL-6 then acted on fetal intestinal epithelial stem cells, rewiring their gene expression machinery in a way that enhanced the development of Th17 cells in the local tissue environment.

Latent Power

For obvious reasons, obtaining evidence that repeated contact with microorganisms alters immune function is more difficult in humans than it is in mice. Nevertheless, using a systems-level approach similar to the one employed in their earlier twin studies, Mark Davis and his colleagues were able to identify effects of latent CMV infection on immune responsiveness in a human population. The results of their analysis of blood immune parameters showed that healthy young adults (20–30 years of age) who were latently infected with CMV displayed multiple indicators of greater overall immune activation than young adults who were not CMV-infected, including elevated serum levels of the cytokine interferon γ. However, the effects of CMV on immune status waned with age and were no longer seen in older adults (60 to more than 89 years of age). In a key experiment, the investigators then vaccinated all the subjects in the study with the seasonal influenza vaccine and measured the levels of anti-influenza virus antibodies induced by the vaccination. They found that latent CMV infection was associated with significantly higher levels of the antibodies in the young adults but not in the older individuals, an effect that correlated nicely with the indicators of immune activation.

The results suggested, but did not conclusively prove, that latent CMV infection had a temporary enhancing effect on immune responsiveness. To establish an actual causal connection, the researchers had to turn to the mouse. They infected a group of lab mice with the rodent equivalent of CMV while another group was left uninfected. At different times afterward, both groups of mice were infected with live influenza virus through the respiratory tract, and the outcome of the infection was determined by measuring the numbers of influenza virus recovered from the lungs (an indicator of how well the infection was controlled) and the vigor of the T cell response directed against the virus. In keeping with the results of the analysis in humans, the mice that were pre-infected with CMV showed stronger T cell responses and better control of the influenza virus than the mice that had not been exposed to CMV. Also like the observations in human subjects, these effects declined with time and were not seen by about three months after the CMV infection.

Innate Cells Can Remember Too

The results of these various experiments, which suggest that previous infections can have effects on the response to subsequent immune challenges, raise some interesting questions. They seem to indicate the operation of memory—prior experience influencing a later response—but everything that we've learned so far about immunological memory implies that it should be microbe-specific. An initial exposure to microbe A will accelerate and enhance the immune response to the same organism if it is encountered again, but it should have no effect on the response to a completely different organism, microbe B. This specificity is what you would expect given what we have learned about the action of memory B and T lymphocytes. So how can infection of mice with viruses or parasites confer enhanced resistance to *Listeria*, a bacterial pathogen? Similarly, how can latent CMV infection in humans boost the antibody response to vaccination against the unrelated influenza virus? The relatively brief duration of some of the effects—three months for the heightened protection against influenza virus following infection of mice with CMV—is also not what would be expected for traditional immunological memory mediated by B and T cells.

Several potential explanations have been offered for these rather surprising observations. One possibility is simple cross-reactivity: the antigens of microbes A and B happen to look like each other and have enough structural similarity that the memory lymphocytes generated against A can also be activated by B. This idea is unlikely to explain the effects that we have discussed since the antigens of viruses and parasites, for example, are very different from those of *Listeria*. Even CMV and influenza virus antigens are not similar enough to induce robust cross-reactivity. A memory effect based on cross-reactive lymphocytes is also unlikely to disappear in a few months.

Another possible explanation relates to a recently described phenomenon known as trained immunity (figure 9.1). Trained immunity is a primitive form of memory that involves cells of the innate immune system, particularly macrophages. When macrophages are exposed for a few hours to certain microbes or microbial molecules—the Bacille Calmette-Guérin (BCG) vaccine strain of *Mycobacterium bovis* and some types of complex fungal carbohydrates are the ones typically used in experiments—the cells respond by enhancing their antimicrobial

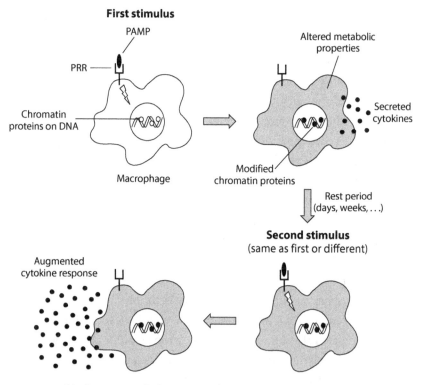

FIGURE 9.1. Mechanisms underlying trained immunity in macrophages.

capabilities and secreting cytokines such as IL-6. In addition, recent experiments indicate that they undergo an extensive reprogramming of function, including alterations of central metabolism and modifications of chromatin proteins that influence the expression of large numbers of genes. If the cells are then subjected to a second stimulus, either the same as before or a completely unrelated one, the reprogramming or "training" induced by the initial microbial exposure leads to a significantly augmented response, manifesting in part as an increase in the amount of cytokines secreted. The augmented response can be observed even if the first and second stimuli are widely separated in time, indicating that the effects of the training are relatively long-lasting. This kind of macrophage-associated innate memory has been observed with isolated cells in tissue culture, in mice and in humans, and has been shown to persist for several months. The characteristics of trained immunity suggest that it could be an important factor in the effects of infection by one pathogen on the response to subsequent encounter with an unrelated microorganism or immune stimulus.

Macrophages are not the only cells of the innate immune system that are capable of remembering: memory-type responses have also been documented in other innate cells such as NK cells. And, as we saw in the experiments on the effects of maternal *Yersinia* infection, even intestinal epithelial cells can be reprogrammed by innate activation so that the response to subsequent immune challenges is altered. Taken together, these observations suggest that a form of immunological memory that is not antigen-specific, something that is present in plants, fruit flies, and worms, may have been preserved through evolution because of the advantage that it confers: having the immune system stimulated into a temporary state of high alert by an initial infection could make it better equipped to deal with another one, even if the same pathogen is not involved.

If you think carefully about the various findings and ideas that we have discussed in this chapter, as well as in earlier ones, they raise the intriguing possibility that each of us may have an immune system that

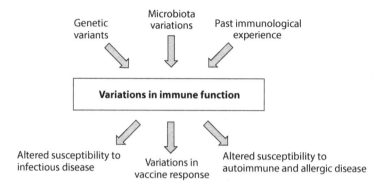

FIGURE 9.2. The factors that can influence immune responsiveness and susceptibility to infection and immune-mediated disease.

is unique, its specific capabilities influenced by the genes that we inherit and the microbiota that we harbor but also shaped by our individual (and even by our mothers') immunological experiences. Every time we suffer through a bad cold or a nasty bout of gastroenteritis, our immune cells are subtly altered, imprinted by changes that may determine the way we respond to the next pathogen or foreign antigen that comes our way. On top of this general tweaking of immune function, some aspects of which may be relatively transient, the memory B and T lymphocytes and the long-lived plasma cells that are generated with each infection or vaccination provide an additional layer of more durable antigen-specific responsiveness (perhaps with a degree of cross-reactivity in some cases) that can last for years and that adds to the competence of our immune systems. Thus, the persistent traces of our past microbial encounters, together with the effects of our genes and our microbiota, could help to explain why I may have asthma while you may have a peanut allergy, or why your response to the flu vaccine may be better than mine (figure 9.2).

CHAPTER 10

Harnessing the Function of the Immune System

The acquisition of knowledge is a worthy and gratifying end in itself. But it is particularly rewarding when that knowledge can be put to practical use. From its inception, the field of immunology has worked toward both goals, with basic discoveries providing the foundation for disease-related applications, and clinical problems spurring the quest for fundamental understanding in a steady, leap-frogging advance. Emil von Behring, who is considered one of the founding figures of immunology, started his research career in the 1890s by working together with Shibasaburo Kitasato on antitoxins, antibodies that neutralize the effects of bacterial toxins. Their findings led to the use of sera enriched in antitoxin antibodies for the life-saving treatment of diphtheria and tetanus (and a Nobel Prize for von Behring but, puzzlingly, not for Kitasato) and also paved the way for the later isolation and structural characterization of antibodies.

Although the basic science and clinical translation of immunology continued to progress hand in hand, research in the field during much of the twentieth century was focused on clarifying in ever greater detail the serological, the cellular, and finally the molecular mechanisms of the immune system. From about the last quarter of the century onward, however, investigators started to direct their efforts increasingly toward finding concrete applications for all the insights that had been obtained and that continued to emerge. The shift in emphasis reflected a natural maturation of the field, but it was also probably driven by funding agencies, which usually have a vested interest in solving real-world problems, and by entrepreneurs, who are keen on finding ways to convert basic scientific discoveries into useful (and marketable) products. Regardless of the exact nature of the forces involved, it is indisputable that a better understanding of the immune system has led to the development of novel strategies for diagnosing, treating, or preventing a number of diseases that would have otherwise taken a far greater and deadlier toll.

Antibodies as Detectors

The exquisite specificity of the antigen-antibody interaction provides the basis for many diagnostic assays that detect clinically relevant molecules. Some of these techniques are used commonly enough that they are probably familiar from personal experience. The home pregnancy test and the rapid strep test are good examples of such ubiquitous antibody-based methods. Both tests make use of specific antibodies to detect the molecule of interest in the appropriate biological sample, a pregnancy-associated hormone present in the urine in the former and a cell wall carbohydrate of *Streptococcus pyogenes* found in pharyngeal secretions in the latter. The tests are designed so that the binding of the antibody to its target antigen leads to the appearance of a visible marker, usually a colored line, which indicates that the relevant molecule is present in the sample. The fact that the antibody in each case binds exclusively to either the hormone or the *S. pyogenes* carbohydrate is the reason that it can be used to identify pregnancy or streptococcal infection with a high degree of confidence. Many other antibody-based antigen detection tests are designed along similar lines and are used to diagnose a variety of infections, including hepatitis B, HIV, and more recently SARS-CoV-2. All these tests depend on the availability of antibodies that bind specifically to the molecule of interest. Such antibodies can be generated by immunizing animals with the relevant molecule or by monoclonal hybridoma technology (as discussed in the next section).

Assays based on antigen-antibody interactions can also be configured to detect the antibodies, rather than the antigens, that are characteristic of infectious diseases. Infections lead to the induction of circulating antibodies against antigenic molecules expressed by the pathogen, IgM initially, IgG in the later stages and during convalescence. The presence of these antibodies in a patient's serum is thus strong evidence that he or she has been infected by the pathogen, with IgM indicating a recent or ongoing infection and IgG an infection sometime in the past. The antibodies can be identified based on their ability to bind to purified antigens that are specific to the relevant pathogen, with the binding acting as the starting point for a measurable read-out such as a color change. Many of these assays allow the quantitation of the levels of pathogen-specific IgM and IgG, providing a way to follow the changes in the antibody response over time. Measurement of pathogen-specific antibody levels by means of such assays is an important way to assess the efficacy of vaccines.

An Army of One: The Development of Monoclonal Antibodies

The early versions of antigen detection assays made use of antibodies purified from the serum of animals, typically rabbits, goats, or horses, that had been immunized with the molecule of interest. These antibodies are polyclonal in nature since they are derived from the activation, clonal expansion, and differentiation of multiple B cells, each of which expresses a unique BCR that recognizes a different part of the antigenic molecule. The production of polyclonal antibodies is a cumbersome process. In addition to requiring careful husbandry, the immunized animals can "donate" only a relatively small volume of blood at any one time and have a finite life span, so the total amount of antibodies that can be obtained is restricted. These limitations were overcome when César Milstein and Georges Köhler developed a radically different method for producing antibodies.

Milstein, who started working at the United Kingdom's Medical Research Council Laboratory of Molecular Biology in Cambridge in the 1960s, was looking for a way to generate antibodies against known antigens in quantities large enough to allow characterization of their structure. Köhler joined the effort in 1974 as a young postdoctoral fellow. He and Milstein soon came up with a clever method for getting what they wanted by building on the work of several other scientists who had collectively generated a wealth of useful information, techniques, and reagents. Their idea was to isolate spleen cells from a mouse that had been immunized with a source of foreign antigens (sheep red blood cells—RBCs—in their first experiment) and incubate them with mouse myeloma cells under culture conditions that promoted the fusion of one cell with another. They also manipulated the ingredients in the culture medium so that only fused, or hybrid, cells would be able to grow. Myeloma cells are cancerous plasma cells that secrete antibodies of unknown antigen specificity and, unlike normal B lymphocytes or plasma cells, they can multiply indefinitely in a lab dish. The spleen cells from the immunized mouse included B lymphocytes activated by the sheep RBC antigens. Milstein and Köhler hoped that among the hybrids that grew up under the conditions of their incubation mix there would be some that were the result of fusion between a myeloma cell and an individual B cell activated by the sheep RBCs. They expected that such "hybridomas" would have the characteristics of both fusion partners—perpetual growth (conferred by the myeloma cell) and the ability to produce a single type of antibody, a monoclonal antibody, against a specific sheep RBC antigen (conferred by the activated B lymphocyte).

After several attempts, the method ultimately worked even better than expected, so much so that Köhler was reported to have gone into a frenzy of shouting and kissing (just his wife, not everyone indiscriminately) when he saw the results during a late-night visit to the lab. Importantly, it was possible to isolate individual hybridomas that each secreted a monoclonal antibody with the ability to recognize a single antigenic region (epitope) on sheep RBCs. Different hybridomas produced monoclonal antibodies against different epitopes. Since hybridomas could be grown indefinitely, the antibodies that they secreted could be purified from the cell culture medium in unlimited quantities. The ability to produce milligram or greater amounts of antibody against a defined antigen of interest represented a significant advance for structural studies. Its importance for the practical application of antibody-based technologies was appreciated only gradually, but even by the time Milstein and Köhler received their Nobel Prize in 1984, monoclonal antibodies generated from hybridomas had found a wide variety of uses, including in many of the diagnostic assays discussed earlier.

Take Two Aspirin . . .

As was discussed in earlier chapters, the inflammatory response is an important component of innate antimicrobial defense. But inflammation can also lead to tissue damage and clinical problems if it persists without resolution. Such harmful chronic or recurrent inflammation is a characteristic feature of diseases like rheumatoid arthritis, lupus, and IBD. An important strategy to manage these conditions, sometimes the only option available if the underlying cause cannot be specifically treated, is to reduce the inflammation.

A number of drugs to inhibit inflammation have been developed over the years. One of the oldest anti-inflammatory agents is aspirin, which has been used in crude form as preparations of willow or meadowsweet flower from the time of the ancient Sumerians and Egyptians. The modern version—acetylsalicylic acid—was synthesized in the mid-1800s and has been in continuous use since then. Its mechanism of action was worked out in the 1970s by John Vane of the University of London, who showed that it inhibited the production of prostaglandins, an important class of eicosanoid lipids that contribute to the inflammatory response. Other nonsteroidal anti-inflammatory drugs (NSAIDs) like ibuprofen and indomethacin work by a similar mechanism. As we have all experienced, aspirin and the NSAIDs are helpful in relieving the inflammatory symptoms associated with common and garden-type infections, musculoskeletal injuries, and the milder forms of arthritic diseases. Vane's studies on prostaglandins and their role in various biologic processes, including the effects of aspirin, earned him the Nobel Prize in Physiology or Medicine in 1982.

Corticosteroids (prednisone, dexamethasone) represent another type of broadly acting anti-inflammatory agent and are more potent than aspirin and the NSAIDs. Steroids also act by inhibiting eicosanoid production, but in addition they suppress expression of several cytokines that are involved in the inflammatory response. Their multiple actions make them very effective at controlling the more severe forms of inflammation that can be seen in IBD, rheumatoid arthritis, and other autoimmune diseases. They are often employed as the initial treatment in these conditions, especially if the inflammation is significant and requires rapid alleviation. However, they have a number of potentially harmful side effects, the result of immunomodulatory and other mechanisms, that significantly limit their long-term usefulness.

Because of the problems associated with nonspecific inhibitors of inflammation like the corticosteroids, vigorous efforts have been made to identify more narrowly acting agents that interfere with specific aspects of the dysfunctional immunity associated with chronic autoimmune diseases. Small molecule inhibitors of T cell activation and proliferation like cyclosporine and tacrolimus are a couple of the results of these efforts, and they have proven useful in the treatment of some forms of chronic inflammation.

Block and Tackle: The Therapeutic Use of Antibodies

More recently, the search for a magic bullet to stop the inflammatory response in its tracks has turned to antibodies. In addition to their application in diagnostic methods, the specificity of antigen recognition makes antibodies ideal for blocking key molecules involved in disease-causing biological processes. We have already learned about the early implementation of this idea in the treatment of diphtheria and tetanus based on the use of sera containing neutralizing antibodies to the corresponding toxins. A similar approach was subsequently employed to prevent the development of other infectious diseases—hepatitis B, for example—by administering appropriate neutralizing antibody preparations to individuals exposed to the pathogen. The success of these methods paved the way for the development of antibody-based strategies for the treatment of noninfectious disorders.

One of the first clinical tests of this type of application was carried out in 1992 by two physician-scientists working at the Charing Cross Hospital in London: Marc Feldmann, an immunologist, and Ravinder Maini, a rheumatologist. It was based on Feldmann's work characterizing the cytokines expressed by cultured cells derived from the inflamed joint tissue of patients with rheumatoid arthritis, a chronic autoimmune disease that was often difficult to control with the drugs then available. He found increased levels of multiple cytokines involved in inflammation, including IL-1 and IL-6, as well as TNFα. Surprisingly, when he added an antibody that neutralized the action of TNFα to the cultures, the expression of all the other inflammatory cytokines decreased, suggesting that TNFα was the key driver of the inflammatory process.

Armed with this information, Feldmann and Maini recruited a small cohort of patients with rheumatoid arthritis who were not doing well on the treatment they were on and injected them intravenously with a neutralizing monoclonal antibody to TNFα. The antibody had been made using the Milstein and Köhler procedure, so was of mouse origin, but it had been "humanized" for clinical use by fusing the antigen-binding region to the constant region of a human antibody. Following injection of the antibody, almost all the patients experienced rapid relief of their symptoms and a reduction in the levels of inflammatory markers. Although the symptoms and inflammation recurred after a few months, they could be alleviated and kept under control by periodic administration of the antibody.

Following this initial success, TNFα blocking therapy, using antibodies or other types of inhibitory biological reagents, has been applied to several chronic inflammatory diseases in addition to rheumatoid arthritis, including other types of joint inflammation, ankylosing spondylitis, a deforming inflammatory process of the spine, the inflammation associated with a scaling condition of the skin known as psoriasis, and IBD. These inhibitors are now being used by millions of people worldwide, generating sales that are predicted to be over $100 billion by 2025. Monoclonal antibodies that block the action of other cytokines have also been developed and are used for the treatment of various inflammatory diseases. Notably, antibodies against IL-17 and the IL-6 receptor have found applications in psoriasis and some forms of arthritis, respectively. In addition, an antagonist of the IL-1 receptor has been used to treat the inflammation associated with gout, rheumatoid arthritis, and a number of other conditions.

Cytokines are not the only molecules that are targeted therapeutically in chronic inflammatory disorders. An important aspect of inflammation is the recruitment of white blood cells into the diseased tissue. As we have seen in an earlier chapter, this movement is initiated following interactions between adhesion molecules expressed on the surface of endothelial cells lining blood vessels in the inflamed tissue and corresponding receptors on the surface of circulating neutrophils, monocytes, and lymphocytes. Studies in mouse models of IBD have identified the specific adhesins and receptors that are involved in recruitment of these cells to the inflamed intestine, and a monoclonal antibody directed against one of these molecules, the $\alpha4\beta7$ integrin, has been used successfully in the treatment of IBD.

Suppression Can Be Risky

Tinkering with the function of the immune system has its drawbacks, and both biologic and nonbiologic anti-inflammatory and immunosuppressive reagents can have side effects that can complicate their long-term use. Most importantly, the inhibition of inflammation and other aspects of the immune response can compromise the ability to contain microbial pathogens. The resulting increase in susceptibility to infection can be a significant problem with corticosteroids and biologic immunomodulators like TNFα blockers. Consequently, patients who are being treated with these reagents have to be screened regularly for the occurrence of candidiasis, tuberculosis, and other types of infections. Since the immune system plays an important role in detecting and eliminating malignant cells, patients on immunosuppressive therapy also have to be monitored for the emergence of certain cancers such as lymphomas.

Controlling the Immune Response during Transplantation

Transplantation is a life-saving procedure for individuals who are suffering the effects of an irreversibly failing kidney, liver, heart, or other vital organ, or who need to have their bone marrow replaced to correct an inherited or chemotherapy-induced deficiency of blood cells. The problem, however, is that the immune system of the transplant recipient will almost always identify the donated organ or tissue as being foreign and will mount an attack that will ultimately destroy or reject it.

What is it that makes the transplanted tissue a target? In principle, it can be any molecule expressed by donor cells that is structurally a little different from the recipient's. Although the vast majority of biological molecules are essentially identical within the human species, there are some that vary in structure from one individual to the next. Of these, the most variable or polymorphic are the class I and class II MHC molecules, which, as we learned in an earlier chapter, are involved in presenting antigenic peptides to T lymphocytes. The extensive polymorphism of MHC molecules across the human population is intrinsic to their function: amino acid variations in the floor and sides of the peptide-binding groove allow the binding of a large number of different antigenic peptides, ensuring that most antigens that are likely to be encountered can be presented for T cell recognition. But this property also means that the MHC molecules are the ones that differ the most between individuals, making them the major antigens targeted during transplantation.

As an example, consider a situation in which individual X receives a transplant from individual Y. X's immune system will include B and T lymphocytes with the ability to recognize Y's MHC molecules since such cells would not have been weeded out by negative selection. So, if X received a kidney from Y, these Y MHC-reactive B and T cells will be activated, and the resulting immune response will lead to the death of the donor organ (the second signal required for the activation of the relevant T cells in X is thought to be provided by the effects of DAMPs released from cells damaged during the transplant surgery). In the case of bone marrow transplantation, there is an additional problem: the T and B cells that develop from the transplanted marrow may attack the recipient's cells, causing a clinical syndrome known as graft-versus-host disease, which is characterized by inflammatory manifestations in the skin, gastrointestinal tract, and liver.

How is it possible, then, that transplantation can be carried out successfully? There are two main strategies that are used to facilitate acceptance of the donor organ. The first is to match MHC molecules between donor and recipient. Each individual has a complement of six different class I MHC molecules and six different class II molecules. Studies of kidney transplantation have shown that the greater the number of MHC matches, particularly with regard to class II, the better the chances of the donor organ surviving in the recipient. MHC matching is feasible in kidney transplantation since the recipient can be kept alive by dialysis until a donor with an optimal match is identified. Usually, bone marrow transplantation can also be delayed until a suitable donor is found. However, matching is usually not possible in transplantation of hearts, lungs, and livers because the clinical status of the patient is often dire and devices to replace these organs are either not available or can be used only temporarily.

The second and more important strategy to facilitate successful transplantation is to suppress the recipient's immune system so that its ability to reject the donor tissue is weakened. Immunosuppression is essential even in situations like kidney transplantation where it might be possible to achieve some degree of MHC matching between recipient and donor. The only circumstance where suppression of the immune system is not required is in transplantation between identical twins.

Several types of immunosuppressants are used during transplantation, often in combination, including corticosteroids, small molecule inhibitors of lymphocyte proliferation like cyclosporine and tacrolimus, and antibodies that deplete T cells. The importance of immunosuppression can be appreciated from the observation that the success rate of transplantation increased dramatically following the introduction of cyclosporine in the 1970s. Unfortunately, the immunosuppressant drugs have to be continued life-long after transplantation and their use is inevitably associated with the risk of opportunistic infections and malignancies. Because of these problems, a good deal of effort is being invested in the development of strategies to induce tolerance to donor MHC molecules in the recipient so that the need for immunosuppression is eliminated or at least reduced. Currently, such approaches remain experimental, although some successes have been reported in clinical trials.

Blood Will Out

Blood transfusion can be considered to be a form of transplantation. But it is unlike transplantation in one important respect, namely, that the bulk of whole blood consists of a cell type—the RBC—that does not express MHC molecules. However, RBCs *do* differ between individuals. This difference was first brought to light in the early 1900s by the Austrian physician Karl Landsteiner, who noticed that RBCs would clump together or agglutinate when blood samples from some, but not all, pairs of individuals were mixed. Based on Landsteiner's experiments and subsequent studies, people can be classified into four major blood groups that reflect differences in the kinds of carbohydrate molecules or glycans that are expressed on the RBC surface. The two relevant glycans (also known as blood group antigens) have been dubbed A and B. Individuals who express one or the other are classified into the blood groups A and B, respectively; those who express both belong to blood group AB; while those who have neither are of blood group O. In addition, the RBCs of most people express a protein known as the Rh antigen, whereas a minority lack this protein, a distinction that divides humans into Rh-positive and Rh-negative blood types.

The human microbiota includes organisms that express carbohydrate molecules that are structurally similar to the A and B blood group glycans. Accordingly, if you are a group B individual, you will develop what are called "natural" antibodies against the A glycan because of stimulation of your anti-A B cells by the cross-reactive microbiota molecules ("natural" because the antibodies are produced without overt exposure to the antigen). However, you will not develop antibodies against the B glycan because B-reactive B lymphocytes would have been eliminated from your repertoire by negative selection. Similarly, if your blood group is A, you will develop antibodies against the B but not the A antigen, and if you are blood group O, you will develop antibodies against both A and B. If you happen to be of the AB group, then you will not develop antibodies against either glycan. If a blood group A person is transfused with B group blood, the preexisting anti-B antibodies in the recipient will immediately bind to the donor RBCs and bring about their destruction. The resulting breakdown of millions of RBCs liberates large amounts of hemoglobin and other intracellular contents into the circulation, leading to a dangerous, potentially life-threatening condition known as a transfusion reaction. Fortunately, it is easy nowadays to prevent this problem by determining the blood groups of donor and recipient to ensure that they match before carrying out a transfusion. As an additional precaution, a bit of the recipient's serum is mixed with the donor blood to confirm that there are no other antibodies that might bind to the RBCs and cause a transfusion reaction, a procedure known as a cross-match.

Adverse reactions can also result if an Rh-negative individual receives Rh-positive blood, so determination of the Rh type is required as part of the screening process before transfusion. However, unlike the anti-A and anti-B antibodies, antibodies against the Rh antigen do not develop naturally. They are induced only if an Rh-negative individual is exposed somehow to Rh antigen-expressing cells, for instance, during a pregnancy involving an Rh-negative mother and an Rh-positive child.

Checkmating Cancer

In contrast to chronic inflammatory disorders and transplantation, where dampening of the immune response is the therapeutic goal, there are some situations where assisting or enhancing the immune response may be beneficial. This concept has been applied most successfully in tumor immunotherapy. Tumor immunotherapy can take several forms. Besides the administration of humanized monoclonal antibodies that block the action of growth factors necessary for the survival and proliferation of tumors, monoclonals directed against molecules that are expressed selectively on the surface of malignant cells can be used to recruit immune cells to the tumor and bring about its demise by ADCC and other mechanisms. The tumor-directed antibody can also be coupled to toxins or radioactive chemicals so that it acts like a guided missile that directly destroys the cancer.

The most recent form of tumor immunotherapy, known as checkpoint inhibition, is based on a very different idea. Rather than directing antibodies against the malignant cells, they are aimed at the T lymphocytes that surround and infiltrate some types of tumor. These lymphocytes should be capable of killing the tumor but appear to be held back from carrying out that function because they are restrained by their own regulatory mechanisms. The antibodies that are used for checkpoint inhibition are designed to free the T cells of that restraint.

Checkpoint inhibitor therapy has its origins in the 1990s, when work from James Allison's lab, then at the University of California, Berkeley, and from Tasuku Honjo's lab at Kyoto University in Japan identified two proteins, CTLA4 and PD-1, respectively, that acted as brakes or checkpoints on T cell activation and function. As we learned in chapter 6, these brakes are turned on as a result of engagement by specific proteins expressed on cells in the local environment of the T cells. In a key experiment, Allison and his colleagues showed that if tumor-bearing mice were injected with an antibody that blocked the action of CTLA4, the brake was released, resulting in the killing of the tumor by the unleashed action of the T cells. Honjo and his

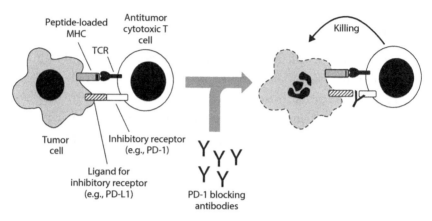

FIGURE 10.1. The principle of checkpoint inhibitor immunotherapy.

colleagues made similar observations with PD-1 (figure 10.1), with the interesting twist added by other investigators that certain tumors expressed high levels of a protein called PD-L1 on the surface of their cells. PD-L1 interacted with PD-1 on T cells to turn on its braking function. In effect, these tumors were able to survive because they were using their own PD-L1 to prevent T cells from killing them.

Based on the mouse experiments, clinical trials were carried out to test the effects of humanized monoclonal antibodies to CTLA4 and PD-1 in several types of human malignancies. Remarkable successes were recorded in subsets of patients with melanoma, renal cancer, and some types of lung cancer. The results were even more striking when combined therapy with both anti-CTLA4 and anti-PD-1 antibodies was used. Patients who previously had little hope that their cancer would regress and expected to live only a few months were blessed almost miraculously with long-term, tumor-free survival. It was a radical improvement in outcome and provided the basis for the Nobel Prize that Allison and Honjo shared in 2018.

Not all types of malignancies respond to checkpoint inhibition. The reason for the resistance of certain cancers varies with the type and is not known with certainty in all cases. Factors that have been implicated include the absence of a significant number of tumor-infiltrating

T lymphocytes (you cannot make use of these cells if they are not there) and the characteristics of other components of the tumor microenvironment. Unfortunately, it is also true that not all patients who have a type of tumor that is known to be responsive to checkpoint inhibition are helped by this form of therapy. This interindividual variability in response is being actively investigated, with some early experiments suggesting that differences in the gut microbiota between responders and nonresponders may be to blame. Finding actionable explanations for nonresponsiveness will help in identifying patients who are most likely to benefit from checkpoint inhibition and in developing workarounds for individuals who do not experience improvement.

In addition to the possibility of treatment failure, the use of checkpoint inhibitors carries a risk of immune-related adverse events that can be serious in a small fraction of individuals. The common feature of most of the adverse effects is uncontrolled inflammation. This problem may occur in the lung, liver, or colon and is not entirely unexpected given that the treatment interferes with a mechanism that normally restrains the immune response and keeps it from getting out of control. CTLA4 and PD-1 are expressed by T cells for a reason.

Tweaking T Cells to Take on Tumors

As revolutionary as checkpoint inhibition is, there is a form of tumor immunotherapy that is even more daring and sophisticated, like something out of science fiction. Based on ideas and techniques that were developed in the late 1980s and then gradually refined for clinical implementation over the course of two decades, it is known as chimeric antigen receptor (CAR)-T cell therapy and involves arming the patient's own T lymphocytes with the means to attack his or her tumor. Several steps are involved in this high-tech process. First, white cells are collected from the patient's peripheral blood, the T lymphocytes are purified, and they are induced to undergo proliferation in the lab using various growth factors and other stimuli. The purified and expanded T cells are then genetically modified by infecting them with a

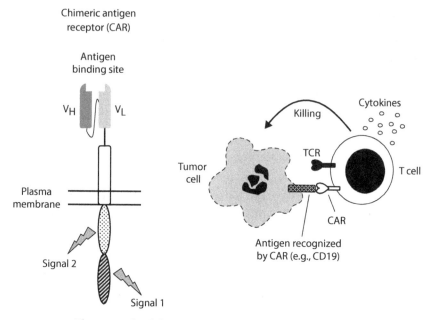

Chimeric antigen
receptor (CAR)

Antigen
binding site

V_H V_L

Cytokines

Killing

Plasma
membrane

TCR

Tumor
cell

T cell

Signal 2

CAR

Antigen recognized
by CAR (e.g., CD19)

Signal 1

FIGURE 10.2. The principle of chimeric antigen receptor (CAR)-T cell therapy.

harmless virus containing a piece of recombinant DNA encoding a CAR, which is an engineered protein that inserts itself into the plasma membrane of the cell and confers the ability to recognize the tumor (figure 10.2).

The CAR has two key parts: the antigen recognition domain, which faces the external environment of the T cell, and a cytoplasmic domain, which faces the inside of the cell. The antigen recognition domain is designed to bind to a specific molecule on the surface of the tumor cells, preferably one that is not found on normal cells, and is made from the variable regions of the heavy and light chains of a monoclonal antibody that is directed against that molecule. The cytoplasmic domain acts as a transducer of signals that are activated when the antigen recognition domain binds to its target molecule on the tumor cell. Very cleverly, the cytoplasmic domain is constructed to have two sub-domains, one that delivers Signal 1 and the other Signal 2, so that together they provide both of the signals required for T cell activation and proliferation. Following the successful introduction and expression of the CAR, the T cells are expanded further to obtain a sufficient number and then infused back into the patient. When a tumor cell is encountered, the binding of the CAR to its target induces the CAR-T cell to proliferate and to kill the tumor cell, either by direct cytotoxic action or by secreting cytokines that recruit and activate other immune cells.

CAR-T cell therapy has been most successful in certain malignancies involving B lymphocytes, either leukemias or lymphomas, and has resulted in about 80% of patients undergoing complete remission. This is a significant achievement considering that many of these individuals have not improved after other forms of therapy or have relapsed after an initial response to treatment. Like most anticancer treatments, CAR-T cell therapy is associated with significant side effects. When used in B cell leukemias and lymphomas, the typical target of the CAR is CD19, a surface protein that is expressed on both malignant and normal B lymphocytes. As a result, the regression of the malignant cells following CAR-T cell infusion is often accompanied by the destruction of normal B cells too, leading to an immunodeficiency state caused by a decrease in antibody levels. The good news is that this problem is usually temporary and can be treated, when required, by administration of purified serum antibody preparations. Other adverse effects include a cytokine release syndrome, which is a manifestation of extensive immune activation induced by CAR-T cells and resembles the effects of a disseminated infection, and a deterioration in cognitive function that has been linked to damage to brain blood vessels. The versatility of the CAR-T cell platform has allowed the introduction of genetically engineered control circuits and other improvements that are directed at reducing the occurrence of these problems. Active attempts are also being made to increase efficacy against solid tumors, which so far have responded poorly or not at all, and to adapt the technology for the treatment of persistent infections and autoimmune diseases. Where CAR-T cells are concerned, it appears that the possibilities are limited only by imagination and creativity.

CHAPTER 11

Vaccination

The Power of Prevention

For sheer numbers of lives saved, there is no immunology-based intervention that has had greater impact than vaccination. Thanks to a vaccine, smallpox, once dreaded as a disfiguring and often fatal disease, has been completely eradicated worldwide. With a bit of luck, we will soon be able to say the same of polio. A recent analysis of the effects of vaccination against ten different pathogens in ninety-eight low- and middle-income countries found that about thirty-seven million deaths were averted just between the years 2000 and 2019, most of them in children below the age of five years. *Thirty-seven million deaths prevented.* That is an astounding figure, equivalent to more than half the current population of the United Kingdom. But the magnitude of that success goes beyond the statistics: think of the children who have had a chance to grow up, of the human potential that has been protected and realized, and you might get some idea of what vaccines have achieved. And we are talking of only twenty years. When you consider that vaccination against smallpox has been practiced for at least half a millennium, against diphtheria and tetanus for over a century, and against several other fatal or incapacitating childhood illnesses for most of our lifetimes, describing the consequences, globally and rippling down the years, would tax the eloquence of a poet. But even in the face of all the evidence of vaccine efficacy and benefit, and of high safety and low cost, there are still many individuals who are hesitant, skeptical, or actively antagonistic about vaccination, a paradox that is distressing and difficult to comprehend.

Edward Jenner Was Not the First

Unlike anti-TNFα biologics, checkpoint inhibition, or CAR-T cell therapy, the initial development of vaccination was not informed by an understanding of immunological mechanisms. The origins of the practice are not clear, but it was probably based on the empirical observation, made as long ago as the fourth or fifth century before the common era, that those who survived smallpox were protected from contracting the disease a second time. A logical extension of the observation was the idea that you might be able to protect against smallpox by deliberately inducing a mild form of the disease in healthy individuals. There is good evidence that this concept was put to use in China and India in the sixteenth century by inoculating individuals with material taken from smallpox lesions, which we now know must have been heavily loaded with the causative virus. There are also hints in the historical record that inoculation against smallpox, described as variolation by Western writers, was a part of religious rituals dating from a much earlier period.

Despite the risks of variolation (about 2% of those inoculated became severely ill and died), its ability to protect against a disfiguring and potentially lethal disease was undoubtedly an important factor in its spread from Asia to the Ottoman Empire, and from there to Africa and Europe. Along the way, it caught the attention of Lady Mary Wortley Montague, an English aristocrat who had accompanied her husband to Constantinople in the early 1700s. She was so impressed that she secretly subjected her son to variolation before returning to England and later spread the word there about the efficacy of the procedure. Her advocacy helped to pave the way for Edward Jenner's famous 1796 experiment in which he inoculated the young James Phipps with material taken from a cowpox lesion (caused by a less virulent relative of the smallpox virus) and showed that the boy was protected against subsequent infection with smallpox.

From the late 1800s onward, starting with Louis Pasteur's work on anthrax and chicken cholera, vaccines against several infectious diseases were systematically developed and analyzed. The early studies were based on the same general idea as variolation, with the important difference that instead of inoculating individuals with the natural, fully virulent pathogen, a harmless or attenuated form of the organism was used to provide protection against disease. These investigations by Pasteur and others focused mainly on devising methods of attenuation rather than on figuring out the mechanisms of protection. However, subsequent research on vaccination and on its laboratory equivalent, the experimental immunization of animals, helped to clarify how vaccines worked and led to many of the important framework concepts of immunology, an example of technology spurring on the basic science.

Making a Good Vaccine

Today, we know that the key characteristic of successful vaccination is the generation of a population of antigen-specific memory lymphocytes, including memory B and T cells and antibody-secreting long-lived plasma cells, which are induced in response to an attenuated form of a pathogen or a pathogen-associated molecule (figure 11.1). The tetanus vaccine, for instance, contains tetanus toxoid, an inactive version of the disease-causing toxin, while the measles vaccine consists of a live but harmless variant of the virus. If the actual pathogen or toxin is subsequently encountered, the preformed antibodies secreted by the long-lived plasma cells provide immediate protection by binding to their target antigens and blocking their harmful effects. At the same time, the memory B and T lymphocytes are activated and differentiate into their respective effector cells, responding quickly and robustly enough that the pathogen or toxin is eliminated before it can do any damage.

Most vaccines have to be administered as a series of initial priming doses before they induce the relevant antibodies and memory lymphocytes in sufficient quantities to confer protection. The levels of vaccine-induced antibodies will usually decline gradually over the

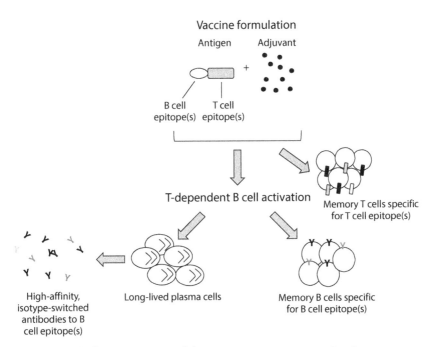

Vaccine formulation

Antigen   Adjuvant

B cell   T cell
epitope(s) epitope(s)

T-dependent B cell activation

Memory T cells specific
for T cell epitope(s)

High-affinity,
isotype-switched
antibodies to B
cell epitope(s)

Long-lived plasma cells

Memory B cells specific
for B cell epitope(s)

FIGURE 11.1. The key components of the immune response to vaccination.

course of years as the long-lived plasma cells reach the end of their life spans—the cells are long-lived, after all, not immortal. Memory lymphocytes may also decrease in number for the same reason. In order to prevent the loss of protection that could result from the attrition of preformed antibody and memory lymphocytes, an additional dose of the vaccine, known as a booster, often has to be delivered. The booster will activate the preexisting memory lymphocytes, eliciting a response that is faster and stronger than the one induced by the priming doses, leading to the generation of fresh populations of long-lived plasma cells and memory lymphocytes that will replenish the ones that are waning. The timing and frequency of the priming and booster doses are usually determined empirically based on the results of clinical studies that monitor responses in large numbers of individuals. So, when and how often you get jabbed depend on the magnitude and effectiveness of the vaccine-induced responses and how long they take to decline, and they can differ from one vaccine to another.

Although memory T cells undoubtedly contribute to the efficacy of vaccination, the protective action of most currently licensed vaccines is mediated to a great extent by memory B lymphocytes and by the high-affinity, isotype-switched antibodies secreted by long-lived plasma cells. The generation of these components of humoral adaptive immunity is optimal when B cell activation occurs in a T-dependent manner, so, if only for that reason, vaccines must induce an appropriate T cell response too. We learned about T-dependent B cell activation earlier (see chapter 5, figure 5.1), but it might be helpful to reiterate some key aspects of the process in the context of vaccination.

T-dependent B cell activation is facilitated by ensuring that the vaccine antigen has two physically linked regions, one (known as the B cell epitope) that can be bound by the BCR of the responding B cell, and the other that is processed into a peptide (the T cell epitope) that can be recognized by the responding T cell when presented on a class II MHC molecule. Binding of the B cell epitope to the BCR induces the B lymphocyte to internalize the antigen and degrade it to generate the T cell epitope, which can then be loaded onto the MHC molecule for cell surface presentation. TCR-mediated recognition of the presented peptide-class II MHC complex allows the relevant T lymphocyte to interact in an antigen-specific manner with the B lymphocyte. The interaction leads to the activation and proliferation of the B cell, and the subsequent production of high-affinity, isotype-switched antibodies to the B cell epitope, as well as the generation of corresponding memory B lymphocytes and long-lived plasma cells. In effect, the linkage of the B and T cell epitopes acts as a message, something like a secret password shared between conspirators, that the two cell types can enter into a productive engagement.

The B cell epitope that is incorporated in the vaccine is chosen so that the antibodies generated against it will interfere with some key disease-causing aspect of the corresponding pathogen or toxin. As an example, the antigen in the SARS-CoV-2 vaccine is the viral spike protein, which is needed for the virus to infect host cells. Vaccination-induced antibodies to the spike, especially if they are of the neutralizing type, will prevent the virus from getting into cells, multiplying, and causing disease.

Importantly, and as we learned earlier, the T cell is capable of activating the B cell only if it has been itself previously activated by a dendritic cell presenting the same peptide-MHC combination as the B cell. This is because B lymphocytes are unable to provide the co-stimulatory signals required for activation of naïve T cells. The requirement that a T lymphocyte must first be activated by a dendritic cell before it can interact productively with a B cell means that a vaccine must include a substance capable of stimulating PRRs. This substance is known as an adjuvant in the context of vaccination or immunization and corresponds to "the immunologist's dirty little secret" mentioned by Charles Janeway. There are various types of adjuvants in clinical use, including aluminum salts, lipid formulations, and derivatives of LPS. They all function like PAMPs or DAMPs to increase expression of the dendritic cell co-stimulatory molecules that provide the second signal needed for T cell activation. They may also have other useful properties such as protecting the vaccine antigen from degradation or facilitating antigen uptake by dendritic cells. The PRR-activating properties of adjuvants and the associated inflammation are what usually cause the unpleasant symptoms—the sore arm, general aches, and fever—that occur sometimes after vaccination.

All the requirements for T-dependent activation of B cells are easily met if the vaccine is made up of whole microbes, either living, attenuated versions, as in the measles or rotavirus vaccines, or killed in some way, as in the influenza vaccine. Multiple, physically linked B and T cell epitopes are present as part of the microbial particle, and the microorganism brings an adjuvant along with it in the form of its own PAMPs. If the antigen is a purified protein, as in the tetanus and hepatitis B vaccines, it must contain both B and T cell epitopes (usually not a problem if the protein is reasonably large), and the vaccine must contain an adjuvant. In the case of a nonprotein antigen, like the bacterial cell wall carbohydrate used in the pneumococcal vaccine, the formulation must have an adjuvant, and the antigen should be physically linked (conjugated) to a "carrier" protein that provides the T cell epitope needed for T-dependent B cell activation.

Mucosal Vaccines: Acting Locally for Global Benefits

Most of the vaccines that we receive as children or in later life are administered by injection into a muscle. They induce B and T cell responses that effectively cover the whole body, including the tissues exposed to the relevant pathogens or pathogenic molecules. In addition, there are a few vaccines that are delivered directly to the sites of pathogen entry, specifically, the mucosal surfaces of the gastrointestinal or respiratory tracts. The rotavirus vaccine, given by mouth, and some types of flu vaccine, administered as a nasal spray, are examples of such mucosal vaccines. The goal of using these routes of administration is to elicit immune responses that are concentrated in the gastrointestinal and respiratory systems, respectively, since these are the entry points for infection by the rotavirus and influenza virus. Moreover, because tissues like the intestine and lung are rich in factors that promote IgA class switching, the antibody responses induced by mucosal vaccines generate large amounts of this isotype. The secretory form of IgA can be transported across epithelial surfaces into intestinal and respiratory secretions, and it can bind to and help to eliminate pathogens even before they have a chance to infect host cells. Mucosal vaccination is thus a particularly effective strategy for preventing diseases caused by pathogens that are transmitted by the inhalation or ingestion of infectious material.

Messengers of Hope

If the COVID-19 pandemic can be considered to have had a silver lining, it was the rapid production of a number of vaccines that made use of both well-established and previously untested strategies to induce protection against the SARS-CoV-2 virus. Developed at unprecedented speed by the coordinated efforts of academic institutions and biotechnology companies across the world, these vaccines have helped to save the lives of millions of people.

One of the most novel anti-SARS-CoV-2 vaccines to be deployed, created independently by the biopharmaceutical companies Pfizer/BioNTech and Moderna, was produced and authorized in record time, thanks in part to the unorthodox technology that it uses. These vaccines do not contain an actual antigen. Rather, they are formulated as millions of tiny nanoparticles, each particle consisting of a lipid coat that surrounds chemically modified mRNA molecules encoding the SARS-CoV-2 spike protein. Following injection into the muscles of the upper arm, the mRNA is carried by its coating of lipid into local cells, including muscle cells, macrophages, and dendritic cells, and is then translated to produce the spike protein. The modifications of the mRNA enhance its stability, facilitate its translation, and also dampen activation of inflammatory responses that might otherwise be excessive and detrimental. Dendritic cells carry the spike protein, either expressed within themselves or picked up from other cells, to the local lymph node, where it is presented to B and T lymphocytes (as peptide-MHC complexes in the case of the latter) to initiate an adaptive immune response. The adjuvant action needed for the vaccine to induce a strong T-dependent B cell response is provided by the lipid coat of the nanoparticle, and possibly also by the residual PRR-stimulating activity of the mRNA. The advantages of the mRNA-based technology include ease and efficiency of production, since it avoids the complex and costly steps involved in purification of proteins, and the flexibility of being able to alter the antigen by simply changing the mRNA sequence.

The development of the Pfizer/BioNTech and Moderna SARS-CoV-2 vaccines would not have been possible without the dedicated (and, until recently, largely unnoticed) efforts of several scientists who worked for years to optimize the use of mRNA for the therapeutic expression of proteins in cells. Katalin Karikó had been intrigued by this idea from the time that she was a graduate student in her native

Hungary and continued to pursue it with dogged determination at the University of Pennsylvania despite a chronic shortage of funds, recognition, and support. In 1997, she and Drew Weissman, a physician-scientist colleague who was trying to develop a vaccine for AIDS, succeeded in delivering and expressing an HIV protein-encoding mRNA in dendritic cells. But Karikó was rather dismayed that the mRNA provoked a strong inflammatory response in the cells. Although the response was potentially helpful for some aspects of vaccination, it could also have deleterious effects on translation of the mRNA and on the generation of T cell immunity.

Karikó and Weissman spent the next several years trying to tame the mRNA-induced inflammation. By comparing several types of RNA, they discovered that they could significantly attenuate the inflammatory effects of mRNA by substituting some of its natural nucleoside building blocks with chemically modified versions. They were elated by this achievement, something that they viewed as a major advance in the field of RNA therapeutics, but it attracted little attention. It was only in 2010, when Derrick Rossi, an investigator at Harvard Medical School, showed that the Karikó-Weissman approach could be used to reprogram stem cells that the full potential of modified mRNA was more widely appreciated. Rossi went on to establish Moderna (a name that pays homage to the modified RNA technology), and in 2013, Karikó joined BioNTech, a German company that had been founded by a pair of Turkish scientists, Ugur Sahin and his wife Özlem Türeci, to develop cancer vaccines using mRNA-based methods. Like Karikó, Sahin and Türeci had spent many years tinkering with mRNA and had identified discrete sequence alterations of the noncoding portions of the molecule that made it more stable and also increased its ability to be translated into protein. With the addition of Karikó and Weissman's modified nucleoside strategy, they had a powerful platform from which to launch a number of applications involving mRNA-directed protein expression.

In January 2020, when the first tremors of the COVID pandemic were being felt, BioNTech's mRNA technology represented what was probably the quickest route—a veritable autobahn—to a successful SARS-CoV-2 vaccine, and Sahin and Türeci immediately redirected the company's considerable resources toward that goal. Moderna also joined the race to develop an mRNA vaccine against COVID. The rest, of course, is history. In 2021, Karikó and Weissman finally received the recognition they deserved in the form of the Lasker-DeBakey Clinical Medical Research Award and the William B. Coley Award, two of the most significant honors in the field of biomedical science. Sahin and Türeci also received the Coley Award and have become the faces of immigrant success in Germany.

Beyond Antigen-Specific Protection

The protection provided by vaccination is usually antigen-specific. Thus, the memory lymphocytes and antibodies induced by the tetanus vaccine, for instance, will not protect against diphtheria. However, some vaccines also confer a degree of nonspecific protection. One of the best examples of this type of effect is provided by BCG, a live, attenuated strain of *Mycobacterium bovis*. BCG is the only available (and rather poorly effective) vaccine against TB, the chronic disease caused by the related mycobacterium *M. tuberculosis*. It is administered routinely to newborns in many countries, and several studies have shown that babies who receive the vaccine at birth are less likely to die of infections in general, including those caused by pathogens completely unrelated to *M. bovis* or *M. tuberculosis*. This nonspecific protective effect is most prominent during the first month of life, although it is still appreciable up to the end of the first year. The measles vaccine also provides a similar nonspecific protection against unrelated pathogens in addition to its outstanding specific efficacy against the measles virus. It is not clear why some vaccines confer nonspecific protection, but trained immunity is one of the mechanisms that has been implicated in the case of BCG. The measles vaccine may protect against other pathogens by preventing the development of the immune amnesia that can follow natural measles infection. Cross-reactivity based on structural similarities between antigens in the vaccine and those expressed by unrelated pathogens is also a possibility but is less likely.

"No Man Is an Island"

The most obvious protective effect of vaccines is on the individual. A child who is vaccinated against measles, for instance, acquires immunity to the virus that is not dependent on the status or actions of others. But in addition, vaccines can provide less apparent, yet equally important, protection at the level of the community. When the majority of individuals in a population are vaccinated and no longer vulnerable to measles, transmission of the virus falls to very low levels and may ultimately cease altogether. Under these circumstances, even those who are unvaccinated are unlikely to become infected since there are simply not enough susceptible people to sustain spread of the virus from one person to the next. This population-based resistance to infection is known as herd immunity. The situation is somewhat similar to that of a neighborhood in which most of the houses have watch dogs. Even though a few of the properties do not have dogs and are not directly guarded, their occupants can sleep easy at night because potential thieves are likely to stay away from the area as a whole. The proportion of the population that must be vaccinated in order to achieve herd immunity depends on how infectious the pathogen is— the more transmissible the pathogen, the higher the percentage of people who will have to get the vaccine. Regardless of the exact numbers, what is important to remember is that when you get vaccinated, you reap benefits not only for yourself but also for those around you.

Although we understand the basic principles that govern vaccine efficacy, vaccinology continues to be an active, fertile, and essential area of research. As the SARS-CoV-2 pandemic has made abundantly clear, emerging pathogens have the ability to bring the world to its knees. They are likely to be recurring problems in coming years and will need to be countered with ingenuity, organization, and cooperation on a global scale. The development of novel technologies for rapidly formulating and equitably distributing vaccines will have to be a part of the response to such widespread outbreaks of infectious disease.

Meanwhile, good vaccines against old scourges like TB, malaria, and AIDS are still lacking. Another particularly vexing issue is that some existing vaccines, like the oral polio vaccine and the rotavirus vaccine, are poorly effective when they are administered to children in low- and middle-income countries, even though they provide excellent protection in other parts of the world. We do not have a satisfactory explanation, nor a corrective, for this variation in efficacy. Finding solutions for all of these problems, and many others like them, will undoubtedly keep immunologists busy for years to come and could have as much impact on future generations as the smallpox vaccine had on previous ones.

CHAPTER 12

"To Follow Knowledge Like a Sinking Star"

I have been an entranced student of immunology for the last thirty years. It has been my privilege to witness at close quarters some of the most exciting developments in the field as it moved from the whole organism, to the cellular, and then to the molecular level. When I first started as a postdoctoral trainee, immunologists had only the haziest idea about the presentation of antigen to T cells. It was known that peptides must be generated and that they must be transported to the cell surface in some way, but how that happened was a mystery. Now, each step of the process has been worked out in exquisite detail, with most of the molecules that play a major role identified and characterized. The mechanism by which T cells recognized antigen was also a subject of much debate and hand-waving. I remember the first puzzled speculations about the nature of the TCR. How could it recognize both antigenic peptides and MHC? Was it just one receptor or were there two—or perhaps even one and a half, as one peculiar model suggested? With the cloning of the TCR, which was reported just before I started my research career, and the follow-up studies that elucidated its structure, its workings were finally revealed in all their beauty, and everything was suddenly made clear.

Recently, the field has shifted again, from cells and molecules back to studies in whole organisms, with an increasing emphasis on human immunology. The shift has been enabled by the relatively facile creation of genetically engineered mice, animals in which individual genes have been knocked out or modified in various ways in order to reveal their contributions to the immune response. At the same time, the use of sophisticated microscopic imaging tools has allowed visualization of the real-time movement and behavior of cells in living tissue, providing an unprecedented view of the immune system in action. Other technical advances, applicable to both mice and humans, have led to increasingly detailed analyses of the microbiota, including its community structure, functional capabilities and metabolites, and of the phenotype, gene expression patterns, and chromatin structure of individual immune cells in a population. The huge data sets that are generated by these techniques can be integrated using novel statistical methods (frighteningly complex for those of us who are math averse) to arrive at correlations and novel hypotheses. The hypotheses can then be tested in mice, including gene knockout strains when required, providing a relatively smooth segue from correlation to causality. This approach has been particularly informative in humans and has helped to clarify how variables such as age, lifestyle, disease, and vaccination influence the immune system.

The exponential increase in knowledge that has occurred over the last few decades may lead some to wonder if we are approaching a plateau in our understanding of the immune system. All the big issues appear to have been resolved, and it seems increasingly unlikely that anything yet to be discovered will fundamentally alter the way we think about immunity. Is there anything really important left to learn? Charles Janeway posed similar questions in a famous paper published in 1989 but then proceeded to describe several immunological puzzles that remained to be tackled, along with his own prediction that innate immunity, which was viewed at that time as something of a bit player, would be the key to their solution. As subsequent discoveries showed, that prediction proved to be startlingly prescient, and innate immune cells have turned out to be at least as important as B and T lymphocytes in executing and regulating the immune response. Although we have come a long way since then, I think we are still far from the twilight era of immunology. There are gaps in comprehension that have to be filled in, disease mechanisms that remain to be clarified, and, certainly, many applications of existing concepts that are waiting to be designed and implemented. Each advance provides a fresh perspective on what lies ahead, and a new generation of Janeways will undoubtedly emerge to point the field in the right direction.

Assimilating all the information that has been and continues to be generated is no easy task. But it is an enjoyable one, accompanied as it is by a sense of things fitting into place. It is similar in some ways to putting together a tricky model airplane or battleship (a hobby that occupied many weekends of my childhood) except that the instructions are minimal and there is no equivalent of the picture on the box to guide assembly. So, what you get at the end—the "Aha!" moment when it all suddenly makes sense—can be particularly satisfying because it is unexpected. Of course, the model often remains half done for years, the sides of the fuselage or hull glued in place, perhaps, but important pieces yet to be found, so that you have only a vague idea of the final form that will appear.

That is probably the current state of play as far as understanding of leprosy is concerned. Individuals who develop tuberculoid leprosy appear to have a strong Th1-type immune response, which helps to control growth of *M. leprae*, but at the expense of early and significant nerve damage. In contrast, patients with lepromatous leprosy have a more Th2-dominated response, which is ill-suited to dealing with an intracellular pathogen and allows unrestrained multiplication of *M. leprae* in cutaneous nodules, its only saving grace being the avoidance of major nerve damage until relatively late in the disease. Despite these important insights, we still don't know why some individuals mount one type of response and not the other, or why most people who are exposed to *M. leprae* do not develop the disease at all. Genetic polymorphisms might account for some of this variation, but they are probably only one part of the story. Nongenetic factors also undoubtedly play a role. What they are and whether they act via differences in the microbiota, previous immunological experience, or something else entirely remain to be discovered. But even knowing that these are the unknowns, the possibilities to be considered, represents an advance.

When I attend seminars these days, I invariably come away amazed, both by the mind-blowing ways in which molecules, cells, and organisms can be manipulated, and by the fact that I am there, partaking vicariously in the thrill of a surprising insight. These are all developments I would have never imagined in the days that I was riding in the White Elephant. Working in science has been frustrating at times— failed experiments, rejected manuscripts, and unfunded grant applications all take their toll—but one of its enduring pleasures has been hearing about the startling discoveries that seem to occur every week. I mentioned this one day to an old friend, who also happens to be a biomedical researcher, and asked if he felt the same way. "Certainly not!" he said quite emphatically, perhaps even a little indignantly. "The only thing I feel is jealousy. Because *I* should be the one at the podium presenting all that spectacular data." He has a point. Do I wish I could have been the one up there? Of course. But after all is said and done, I'm happy that I've had a seat in the audience.

ACKNOWLEDGMENTS

Many people were involved, directly or indirectly, in creating this book, and I am grateful to all of them for their contributions. Its origins can probably be traced back to my father, who spent much of his life studying the minutiae of tRNA structure and passed on his curiosity about the workings of cells and molecules to his children. That curiosity was reinforced in my case by an outstanding group of teachers and students at Christian Medical College, Vellore, India, where I was privileged to receive my medical education. Of the many talented, hardworking, and incredibly dedicated individuals I met there, I would like to pay special tribute to Prof. P. Zachariah, whose eye-opening lectures on human physiology were models of clarity and logic, and to my classmates Subhendra Banerjee, Vishwajit Nimgaonkar, and the late Dhananjay Abhyankar, whose steadfast camaraderie was instrumental in my academic survival and development. I have to thank another close friend from that time, Anindya Dutta, for catalyzing my transition from the clinic to the laboratory.

A good deal of the material in the preceding pages is based on lectures I have delivered every year at two vaccine courses, INDVAC and Afro-ADVAC, held in India and South Africa for over a decade. INDVAC was the brainchild of Anuradha Bose of the Department of Community Health at Christian Medical College. She started the program in Vellore in order to provide training in vaccinology to health professionals in India and other Asian countries. Afro-ADVAC extended the concept to Africa under the direction of Clare Cutland of the Vaccines and Infectious Diseases Analytics (VIDA) Research Unit, University of the Witwatersrand, Johannesburg. I would like to thank Anu and Clare for giving me the opportunity to contribute to these wonderful courses and for inadvertently prompting me to think about converting my talks into a book.

Several people read the manuscript at various stages and assisted in its progress. My wife, Nandini Sengupta, looked over it with a clinician's eye on a long flight to India, and my daughter Monisha provided valuable insights from the perspective of a nonscientist. Many thanks to my brother Binny, who took a break from physical chemistry in order to read and critique something completely different. I would also like to thank Cariappa Annaiah, Subhendra Banerjee, Amitabha Chaudhuri, Anindya Dutta, Nitya Jain, Shiv Pillai, and Haining Shi for their constructive feedback and for many years of friendship and support.

Immunology is a vast and ever-expanding subject, and I would not have been able to convey some of its intricacies without the assistance of experts in its subdisciplines. I am particularly indebted to my colleagues Adam Finn of the University of Bristol, Jonathan Kagan of Boston Children's Hospital and Harvard Medical School, and Peter Turnbaugh of the University of California, San Francisco, for taking the time to read the sections on vaccination, innate immunity, and the microbiota, respectively, and for offering very thoughtful criticism. Their comments and suggestions helped to clarify ideas and correct inaccuracies. Any remaining errors of omission or commission are my own.

I am very grateful to the staff of Johns Hopkins University Press, particularly Will Holmes, Alena Jones, Juliana McCarthy, Jane Medrano, and Joe Rusko, for helping me to negotiate the multiple steps involved in making a book out of a manuscript. Joe deserves special thanks for taking a chance on a first-time author whose previous writing experience was confined to rather dry research articles. I would also like to thank Terri Lee Paulsen for her meticulous copyediting.

Lastly, but most importantly, I would like to express my heartfelt gratitude to my family. Nandini has been the ideal companion for the unplanned journey that we started almost forty years ago, an empathetic fellow traveler who has taken every challenge in stride and has turned obstacles into opportunities. Our daughters, Monisha and Maya, and our granddaughter, Soraya, have made every stage of that journey an adventure and a joy.

ADCC	antibody-dependent cell-mediated cytotoxicity
AIDS	acquired immunodeficiency syndrome
APC	antigen-presenting cell
ATP	adenosine triphosphate
BCG	Bacille Calmette-Guérin
BCR	B cell receptor
CAR	chimeric antigen receptor
CD	cluster of differentiation
CGD	chronic granulomatous disease
CMV	cytomegalovirus
COVID-19	coronavirus disease 2019
CTL	cytotoxic T lymphocyte
DAMP	damage-associated molecular pattern
DNA	deoxyribonucleic acid
EGID	eosinophilic gastrointestinal disease
FMT	fecal microbial transplantation
HIV	human immunodeficiency virus
HLA	human leukocyte antigen
IBD	inflammatory bowel disease
Ig	immunoglobulin
IL	interleukin
ILC	innate lymphoid cell
LPS	lipopolysaccharide
MAMP	microbe-associated molecular pattern

MHC	major histocompatibility complex
mRNA	messenger RNA
MSMD	Mendelian susceptibility to mycobacterial disease
NK	natural killer
NSAID	nonsteroidal anti-inflammatory drug
PAMP	pathogen-associated molecular pattern
PRR	pattern recognition receptor
RAG	recombination activating gene
RBC	red blood cell
RNA	ribonucleic acid
rRNA	ribosomal RNA
SARS-CoV-2	severe acute respiratory syndrome coronavirus 2
SCFA	short-chain fatty acid
SCID	severe combined immunodeficiency
SLE	systemic lupus erythematosus
SPM	specialized pro-resolving mediator
TB	tuberculosis
TCR	T cell receptor
Tfh	T follicular helper cell
Th	T helper cell
TLR	Toll-like receptor
TNF	tumor necrosis factor
Treg	T regulatory cell
tRNA	transfer RNA
XLA	X-linked agammaglobulinemia

adaptive immunity. The phase of the immune response that involves B and T lymphocytes. It takes 1–2 weeks to develop, is exquisitely antigen-specific, and is associated with immunologic memory, a heightened and more rapid response on subsequently encountering the same antigen.

adhesin. A surface molecule involved in adhesive interactions between cells, or between cells and a nonliving surface.

adjuvant. A substance that is used to enhance immune responses during immunization or vaccination. It functions like a PAMP or DAMP by providing the second signal needed for T cell activation, and it may have additional beneficial effects, including on antigen stability and uptake.

aerosol. Tiny aqueous particles that are finely dispersed in the air, during sneezing or coughing, for example. The aerosol can act as a vehicle for transmission of respiratory pathogens such as the influenza virus.

affinity. The tightness of binding of one molecule to another.

agammaglobulinemia. A condition characterized by abnormally low levels or complete absence of circulating antibodies, resulting in susceptibility to certain infections.

allele. Alternate versions or variants of a specific gene.

allergy. An adverse health effect of an inappropriate immune response to a normally harmless environmental antigen. The manifestations include relatively mild ones like hives or a runny nose, as well as more severe problems such as difficulty breathing and inability to maintain circulation to the tissues.

amino acid. The basic chemical building block of proteins. There are twenty naturally occurring amino acids that are used to make proteins.

anaphylaxis. A severe, systemic form of allergy characterized by a marked fall in blood pressure and impaired circulation, sometimes accompanied by narrowing of the airways and difficulty breathing.

anergy. A state of immunologic unresponsiveness, usually referring to B and T lymphocytes.

antibody. The secreted form of the BCR that is produced by plasma cells. Also known as immunoglobulins, antibodies represent the effector molecules of the B cell (humoral) arm of adaptive immunity.

antigen. A molecule that induces an adaptive immune response and that is specifically recognized by the BCR, antibody, or TCR.

antimicrobial peptide. A short chain of amino acids that has the ability to kill or inhibit the growth of microorganisms. A number of different antimicrobial peptides are secreted by various cells of the immune system, including epithelial cells, neutrophils, and macrophages.

ATP. Adenosine triphosphate, the molecule that acts as the energy currency of the cell. It is produced during the breakdown of nutrient molecules such as glucose and provides the energy to fuel other cellular processes.

autoimmunity. A condition in which the immune system mounts a response against self antigens.

bacteriophage. A virus that infects bacteria.

biosynthesis. The production of biological molecules, usually by cellular enzymatic processes.

bone marrow. The tissue within the cavities of bones. It contains large numbers of progenitor cells that give rise to various cellular components of the blood and immune system.

carbohydrate. A major class of biological molecules made of atoms of carbon, hydrogen, and oxygen linked to each other in different configurations. Carbohydrates include relatively simple sugars like glucose, as well as more complex compounds such as dietary fiber.

CD (cluster of differentiation) molecules or antigens. A set of over three hundred cell surface molecules, each identified by a CD number, that can be detected by specific antibodies. The pattern of CD molecules expressed by a cell is often used as an indicator of cell type. T cells, for example, are characterized by expression of CD3 and can be further classified into different subsets based on expression of CD4 or CD8.

chemokines. A family of structurally related secreted proteins that guide the movement of various immune cells.

chromatin proteins. Proteins that are closely associated with genomic DNA in the chromosomes of mammalian cells and that can contribute to the regulation of gene expression. Modification of chromatin proteins by the addition of various chemical groups influences their functions.

chromosome. Organized structure of protein and DNA that represents a storehouse of genetic information. The typical human cell contains twenty-three pairs of chromosomes, with one member of each pair being inherited from the mother and the other from the father.

clone. In the context of the immune system, a clone signifies a group of lymphocytes that have identical antigen receptors and that results from the proliferation of a single activated B or T cell.

complement. A group of proteins that circulate in an inactive state in the blood. They can be activated in a serial cascade by bacteria or by certain types of antibodies following binding to antigen. Activated components of complement have various functions, including amplifying the inflammatory response, facilitating phagocytosis, and direct killing of some types of bacteria.

corticosteroids. A group of structurally related molecules with diverse functions. In the immune system, corticosteroids act as broad anti-inflammatory agents by inhibiting the production of cytokines and eicosanoids.

cytokines. A family of several dozen different proteins that are secreted by cells of the immune system following activation, with each cell type producing a characteristic repertoire of cytokines. The cytokines can have a variety of effects on surrounding cells as a result of binding to specific receptors.

cytokine storm (cytokine release syndrome). A state of excessive and unregulated production of inflammatory cytokines, often occurring during severe infections. It can manifest with low blood pressure, impaired circulation, and compromised function of multiple organs.

cytoplasm. The semifluid material, composed largely of water, proteins, and salts, within the plasma membrane and outside the nucleus of the cell. Subcellular organelles such as the mitochondria and endoplasmic reticulum are found in the cytoplasm.

cytoskeleton. A network of cytoplasmic protein filaments and fibers that functions as the supporting framework or scaffolding of the cell.

cytotoxic activity. An activity that leads to cell damage or death.

DAMP. Damage-associated molecular pattern, a molecule that is normally present only inside cells, for example, ATP or proteins in the nucleus. When such molecules are detected outside the cell, they indicate cell damage and can initiate an innate immune response.

degranulation. The concerted release of contents stored inside cytoplasmic granules. Mast cells, for instance, undergo degranulation during allergic responses.

dendritic cell. A type of innate immune cell present in most tissues that is characterized by diaphanous membrane protrusions. It plays a key role in antigen processing and presentation to T cells.

domain. When used in the context of describing a protein or other complex molecule, the term usually refers to a structural region that has a specific function, topology, or biochemical characteristic.

dysbiosis. Abnormal composition of the microbiota, often in association with specific disease states.

eicosanoid. A broad category of small, functionally active lipids that are involved in the inflammatory response and in other biological processes.

emulsifier. A chemical additive that is used in food processing to stabilize ingredients, particularly mixtures of oily and nonoily substances.

endocytosis. The process by which substances in the environment outside the cell are taken into the cell, specifically, into a membrane-bound compartment or organelle known as the endosome. It may involve the binding of a molecule to a specific receptor, followed by internalization of the receptor into the endosome (receptor-mediated endocytosis), or the uptake of any molecule(s) dissolved in the fluid outside the cell (fluid-phase endocytosis).

endoplasmic reticulum. A membrane-bound compartment or organelle within the cell where proteins and other molecules enter the export pathway that leads to the cell exterior, the plasma membrane, or other organelles.

endothelial cells. The cells that line blood vessels.

enzyme. A molecule, usually a protein, that catalyzes biological reactions.

eosinophil. A type of blood cell characterized by pink-staining cytoplasmic granules.

epidemiology. The study of factors that influence health and disease at the population level.

epinephrine. A hormone made by the adrenal gland that has a number of effects, including increased heart rate, constriction of peripheral blood vessels, and relaxation of airway muscles. These actions make the hormone useful in alleviating severe allergic reactions.

epithelium. The layer of cells that covers the skin and lines the surfaces of the gastrointestinal, respiratory, and genitourinary tracts. The cells that make up the epithelium can be a single layer, as in much of the intestine, or can be several layers thick, as in the skin. Individual epithelial cell types contribute specialized functions to the properties of the epithelium.

epitope. The specific structural region of an antigen that is bound by the BCR, antibody or TCR. A single antigen may consist of multiple epitopes.

eukaryotic. A type of cell in which genomic DNA is contained within a morphologically distinct nucleus that is separated from the cytoplasm by a membrane.

Mammalian, fungal, and protozoan cells are eukaryotic. Bacteria are prokaryotic since they lack a nucleus.

extracellular. The environment outside the cell.

fibroblast. A cell type present as part of the framework of most tissues. It produces some of the noncellular components of tissue such as collagen.

gastrointestinal tract. The organ system, extending from the mouth to the anus, that is involved in ingestion, digestion, and absorption of food, and in the elimination of nondigestible material.

gene. A stretch of DNA that acts as a template for production of RNA. The term is also used to signify a unit of inheritance.

genitourinary tract. The organ systems that are involved in the production and elimination of urine, and in sexual reproduction.

genome. The entire content of DNA-encoded genetic information in a cell. In humans, the genome of each somatic cell consists of the DNA that is packaged in twenty-three pairs of chromosomes in the nucleus and a small chromosome present in mitochondria.

germ cell. The cell type, such as the sperm or ovum, that is generated in the reproductive organs and that provides half of the complement of chromosomes required to form an embryo.

germ-free mice. Mice that are born and raised in the complete absence of microorganisms.

glycan. A complex carbohydrate that is attached to cellular proteins or lipids, usually those that are exported to the cell surface or to the extracellular environment.

histamine. A small molecule released by mast cells that contributes to the allergic response.

homeostasis. The process of self-regulation by which a physiologic system maintains stability in response to changing external conditions.

hybridoma. An immortalized cell created by the fusion of an activated B lymphocyte and a myeloma tumor cell. It secretes a single type of antibody against a specific antigen.

hypha (plural: hyphae). The elongated tube-like structure produced by certain types of fungi.

ILC. Innate lymphoid cell, a type of innate lymphocyte that is found in most tissues and that plays important roles in the early stages of immune defense. Although ILCs resemble B and T lymphocytes in some respects, they do not have antigen receptors. Instead, they respond rapidly, without the need for clonal selection or expansion, to the inflammatory cytokines and other molecules that are released at sites of infection or cell damage. Like T cells, ILCs can be classified into subsets that have characteristic functions, including patterns of cytokine secretion.

immunodeficiency. A condition characterized by the impairment or absence of one or more immune functions. It is usually associated with increased susceptibility to infection.

immunosuppressant. A molecule that inhibits one or more immune functions, usually for therapeutic purposes.

immunotherapy. Any treatment involving the use or manipulation of immune functions.

inflammasome. A large cytoplasmic complex made of multiple proteins. It is activated by PAMPs or DAMPs in the cytoplasm, or by stresses such as disturbances in cellular ion concentrations. An important consequence of inflammasome activation is the maturation and secretion of the inflammatory cytokine IL-1β.

inflammation. Inflammation is a key aspect of innate immunity. The process is initiated by innate immune cells that are activated by PAMPs or DAMPs, and involves dilation and leakiness of local blood vessels, movement of serum proteins and circulating immune cells into the tissue, and stimulation of nerves. These events lead to the cardinal clinical manifestations of inflammation: redness, warmth, swelling, and pain.

innate immunity. The first phase of the immune response, occurring within minutes to hours of the sensing of a PAMP or DAMP. It involves cells present in or recruited to the site of infection or injury, such as epithelial cells, macrophages, dendritic cells, and neutrophils, and provides initial antimicrobial protection. Importantly, it also leads to the activation of adaptive immunity.

insulin. A hormone produced by the pancreas that is involved in controlling blood glucose levels.

interferons. A family of cytokines produced by most cells (type 1 or innate interferons) or by lymphocytes, especially Th1 effector T cells (type 2 interferon, interferon γ). Both types induce the expression of large numbers of genes that contribute to antimicrobial defense, including in protection against viruses (types 1 and 2) and intracellular pathogens (type 2).

interleukins. A large family of proteins, including various cytokines, that are secreted by cells of the immune system and that have diverse effects on other cells.

intracellular. The environment inside cells.

intron. A stretch of nucleosides in the initially generated mRNA (the primary transcript) that is not present in the mature form. In contrast, exons are the nucleoside segments that are present in both the primary transcript and the mature mRNA. During mRNA maturation, the introns are removed and the exons are joined together in a process known as splicing.

ion. An electrically charged atom or molecule. Examples include positively charged calcium or potassium ions and negatively charged chloride ions.

isotypes. Different types of antibodies that are distinguished by relatively minor variations in the structure of the heavy-chain constant region. There are five isotypes: IgM, IgG, IgA, IgE, and IgD.

larynx. Commonly known as the voice box, it is the upper part of the airway that contains the vocal cords.

ligand. A molecule that binds specifically to a receptor.

lipid. A major class of biological molecules that are derived from or made of fatty acids.

lumen. The cavity of a hollow organ such as the intestine.

lymph. Tissue fluid that is collected by lymphatic vessels and ultimately returned to the blood.

lymph node. Collections of B and T lymphocytes, dendritic cells, and other cell types that are located at strategic points around the body. Lymph from specific regions of the body drain into the local lymph nodes through lymphatic vessels.

lymphocyte. A type of immune cell that is characterized by a relatively large nucleus and a small amount of cytoplasm. B and T cells are the two major lymphocyte types and play important roles in adaptive immunity.

lymphoid organ or tissue. Organized collections of lymphocytes and other cells. Examples are lymph nodes and the spleen.

lysosome. An intracellular, membrane-bound compartment or organelle containing enzymes that aid in the degradation of phagocytosed or endocytosed material.

macrophage. A highly phagocytic and vigorously antimicrobial innate immune cell that is found in virtually all tissues. Monocytes in the blood can give rise to macrophages, usually after leaving the circulation, but tissue macrophages also originate from other precursors during development of the embryo.

mast cell. A resident innate immune cell of most tissues characterized by the surface expression of the high-affinity IgE receptor and by cytoplasmic granules containing many bioactive molecules. The coordinated release of the granule contents following triggering of the IgE receptor is an important component of a major type of allergic response.

metabolism. The breakdown and synthesis of biological molecules that provide the energy and chemical intermediates (metabolites) needed for cellular functions.

microbiota. The community of microorganisms resident on the body's skin and mucosal surfaces.

microglia. The resident macrophages of the central nervous system.

micronutrient. A dietary component that is required in relatively small amounts for health and well-being (for example, minerals like iron and zinc and the various vitamins). In contrast, macronutrients like proteins and carbohydrates are required in relatively large quantities.

monoclonal antibody. An antibody that is the product of a single clone of activated B cells. Each monoclonal antibody binds to a single, specific region of its target antigen.

monocyte. A circulating immune cell type that can be recruited into inflamed tissues (or even into some non-inflamed tissues like the gut) to give rise to macrophages.

mucosa. The epithelium and associated tissues that line the lumens of the gastrointestinal, respiratory, and genitourinary tracts.

mucus. A slimy mix of proteins that are heavily modified by various glycans and that are secreted by specialized cells in the epithelium of mucosal tissues.

mutation. An alteration in DNA structure that may or may not have functional consequences. The abnormality may take the form of deletions, insertions, or changes in individual nucleotides.

mycobiome. The fungal component of the microbiota.

myelin. The fatty layer that surrounds the long extensions (axons) of certain types of nerve cells. This layer facilitates conduction of electrical impulses along the axons.

myeloma. A cancer of antibody-secreting plasma cells.

nanoparticle. A microscopic particle, generally 1 to 100 nanometers in diameter. It can be generated naturally from biological molecules or be made from synthetic components.

neoplasm. An abnormal growth of cells manifesting as a tumor. Neoplasms are described as malignant or cancerous when they invade surrounding tissue, often destructively, and spread to distant sites. They are benign if they do not have these characteristics.

neuron. A nerve cell.

neutralization. The ability of an antibody to bind to a pathogenic molecule, either free-floating or associated with a microorganism, such that the function of the molecule is inhibited.

neutrophil. A phagocytic, highly antimicrobial innate immune cell that is a major circulating white blood cell type. It has a characteristic segmented nucleus and granular cytoplasm.

NK cell. Natural killer cell, a type of innate lymphocyte that is found in the circulation and certain tissues and that shares some characteristics with B and T cells. However, NK cells differ from adaptive lymphocytes in being able to deploy their effector functions—killing of infected and malignant cells and cytokine secretion—very rapidly and, in most cases, in an antigen-independent manner. The mechanisms of NK cell activation are complex and involve sets of activating and inhibitory surface receptors that sense the status of surrounding cells.

nucleic acid. An important category of biological molecule that exists in two forms, DNA and RNA. Each is a chain of repeating subunits called nucleotides (in DNA) or nucleosides (in RNA). There are four types of subunits in each form of nucleic acid, and the sequence in which they are arranged encodes genetic information.

opsonization. The coating of particles, including microorganisms, by antibodies or complement components so that they become easier to phagocytose by macrophages and neutrophils.

organelle. A cellular compartment or structure, often delimited by a membrane, that is specialized to carry out a specific function. Examples of organelles include mitochondria and lysosomes.

PAMP. Pathogen-associated molecular pattern, a molecule that is characteristic of pathogens, or of microbes more generally.

pathogen. A microorganism with the ability to cause disease.

peptide. A chain of amino acids arranged in a specific order. Peptides are generally short, about 5–20 amino acids, whereas polypeptides are longer and can contain hundreds of amino acids.

phagocytosis. The process by which a cell engulfs and takes up particles, including microorganisms, from the external environment. The compartment or organelle into which the particle is taken up is called a phagosome.

phenotype. The outward manifestation of genetic information (genotype). Eye, hair, and skin color, for example, are all aspects of phenotype. In the context of the immune system, the term also often refers to the pattern of surface molecules, including CD antigens, that can be used to distinguish different cell types.

plasma cell. The antibody-secreting effector cell into which activated B lymphocytes differentiate.

plasma membrane. The membranous bilayer of lipids and proteins that forms the outer boundary of the cell.

polyclonal. An adaptive immune response involving multiple clones of T or B lymphocytes. The end result of a polyclonal B cell response is the production of polyclonal antibodies, a collection of antibodies that are directed against different antigens or different parts of an antigen.

polymorphism. An alteration in the sequence of a region of DNA, usually relatively minor.

postbiotic. A molecular product of the microbiota that has health benefits.

prebiotic. A dietary component that facilitates the growth of beneficial microorganisms.

probiotic. A live microorganism that is consumed for its health benefits.

proteasome. A large complex of enzymes in the cytoplasm that degrades proteins into short peptides.

protein. A molecule with a specific three-dimensional structure made up of a single polypeptide or multiple polypeptides assembled together.

protozoan (plural: protozoa). Single-celled microorganism with an organized nucleus, distinct subcellular structures, and a plasma membrane, but without a cell wall.

PRR. Pattern recognition receptor, a cellular receptor that senses PAMPs or DAMPs.

receptor. A molecule, usually a protein, that interacts specifically with another molecule, which is often referred to as the receptor's ligand, to initiate or facilitate a cellular function.

recombination. When used with reference to the genome, the term indicates the joining of one segment of DNA with another.

respiratory tract. The organ system involved in gas exchange between the external environment and the blood. It includes the structures of the airway—the larynx, trachea, bronchi, and bronchioles—and the air sacs (alveoli) of the lung.

SCFA. Short-chain fatty acid, a small organic molecule derived from the microbial metabolism of dietary complex carbohydrates such as fiber. Acetic acid, propionic acid, and butyric acid are examples of SCFAs.

sepsis. A systemic disease resulting from severe infection that is associated with increased production of several inflammatory cytokines, low blood pressure, and impaired circulation to the peripheral tissues. If untreated, it can lead to dysfunction of multiple organs and, ultimately, to death.

serum. The liquid portion of blood following clotting. It contains several important molecules involved in the immune response, including antibodies and complement. The study of these molecules, especially antibodies, is known as serology. Plasma is similar except that it refers to the liquid part of blood that has not clotted.

somatic cells. The cells that form the tissues of most of the body. They contain the full complement of chromosomes, in contrast to germ cells, which have only half the number of chromosomes.

stem cell. A cell that is capable of self-renewal and that gives rise to all the other cell types of a tissue.

stroma. The supporting cells and nonliving material of a tissue.

synapse. The structure that forms the junction between one nerve cell and another.

thymus. An organ located behind the breastbone in which T lymphocytes develop.

TLR. Toll-like receptor, a member of a family of structurally related, membrane-bound receptors that are involved in recognizing and responding to specific PAMPs.

tolerance. In the immunological context, the term refers to a state of nonresponsiveness to a specific molecule.

toxoid. An inactivated toxin.

trained immunity. A form of immunological memory involving cells of the innate immune system. It is relatively short-lived (weeks to months) and is not specific to the molecule that originally induced it.

transcription. The process by which the genetic information encoded in the nucleotide sequence of DNA is converted to information encoded in the nucleoside sequence of RNA, the latter often being referred to as the transcript. There are three major types of RNA: messenger RNA (mRNA), which provides the information needed to produce proteins; transfer RNA (tRNA), which helps to decode the information carried by mRNA; and ribosomal RNA (rRNA), which is a structural component of ribosomes and is also involved in making proteins.

translation. The process by which the information encoded in the sequence of mRNA is used to generate a polypeptide, a chain of amino acids linked together in a specific order.

virome. The viral component of the microbiota.

virulence. When used in relation to a microorganism, the term refers to the ability to cause tissue damage and disease.

Apart from the textbooks, most of the citations refer to publicly available websites or articles.

General Textbooks of Immunology
Abbas AA, AH Lichtman, S Pillai. *Cellular and Molecular Immunology*, 10th ed. Elsevier; 2021.

Parham P. *The Immune System.* 5th ed. W. W. Norton; 2021.

Murphy KM, C Weaver, LJ Berg. *Janeway's Immunobiology.* 10th ed. W. W. Norton; 2022.

Chapter 1. Conceptualizing the Immune Response
Clinical manifestations of leprosy: https://www.cdc.gov/leprosy/index.html

History and social impact of leprosy: https://leprosyhistory.org

Chapter 2. Innate Immunity I
"Good fences make good neighbors."
—Robert Frost, "Mending Wall"
https://www.poetryfoundation.org/poems/44266/mending-wall

Sender R, S Fuchs, R Milo. Revised estimates for the number of human and bacteria cells in the body. *PLoS Biol.* 2016;14(8):e1002533. https://www.ncbi.nlm.nih.gov/pmc/articles/PMC4991899

Gordon S. Élie Metchnikoff, the man and the myth. *J Innate Immun.* 2016;8(3):223–227. https://www.ncbi.nlm.nih.gov/pmc/articles/PMC6738810

Chapter 3. Innate Immunity II
Toll-like receptors and innate immunity: https://www.nobelprize.org/prizes/medicine/2011/press-release

Chapter 4. Adaptive Immunity
Schatz DG, D Baltimore. Uncovering the V(D)J recombinase. *Cell.* 2004;116(suppl 2):S103–S106. https:www.sciencedirect.com/science/article/pii/S009286740400042X?via%3Dihub

Chapter 5. B Lymphocytes
Kaufmann SHE. Immunology's coming of age. *Front Immunol*. 2019;(10):684. https://www.ncbi.nlm.nih.gov/pmc/articles/PMC6456699

Chapter 6. T Lymphocytes
Janeway CA Jr. Approaching the asymptote? Evolution and revolution in immunology. *Cold Spring Harb Symp Quant Biol*. 1989;(54):1–13. https://www.jimmunol.org/content/191/9/4475.long

Ralph Steinman and dendritic cells:
Nobel laureate Ralph Steinman dies at 68. The Rockefeller University. November 25, 2011. https://www.rockefeller.edu/news/1816-nobel-laureate-ralph-steinman-dies-at-68

Steenhuysen J, M Nichols. Insight: Nobel winner's last big experiment: himself. Reuters. October 6, 2011. https://www.reuters.com/article/us-nobel-medicine-experiment/insight-nobel-winners-last-big-experiment-himself-idUSTRE7956CN20111006

Chapter 7. Immune Dysfunction
Blakemore E. David Vetter was "the boy in the bubble." His short life provided insights into how the rare disorder SCID works. *Washington Post*. January 25, 2020. https://www.washingtonpost.com/health/david-vetter-was-the-boy-in-the-bubble-his-short-life-provided-insights-into-how-the-rare-disorder-scid-works/2020/01/24/b698e774-3d3a-11ea-baca-eb7ace0a3455_story.html

Youness A, CH Miquel, JC Guéry. Escape from X chromosome inactivation and the female predominance in autoimmune diseases. *Int J Mol Sci*. 2021;22(3):1114. https://www.ncbi.nlm.nih.gov/pmc/articles/PMC7865432

"The fault, dear Brutus, is not in our stars,
But in ourselves, that we are underlings."
—William Shakespeare, "Julius Caesar," Act 1, Scene 2
http://shakespeare.mit.edu/julius_caesar/julius_caesar.1.2.html

Soldan SS, PM Lieberman. Epstein-Barr virus and multiple sclerosis. *Nat Rev Microbiol*. 2023;21(1):51–64. https://www.ncbi.nlm.nih.gov/pmc/articles/PMC9362539

Chapter 8. Conditioning of the Immune System by the Microbiota
Milestones in human microbiota research: https://www.nature.com/immersive/d42859-019-00041-z/index.html

Effects of the microbiota on health and disease: https://www.gutmicrobiotaforhealth.com

Sharon G, NJ Cruz, DW Kang, et al. Human gut microbiota from autism spectrum disorder promote behavioral symptoms in mice. *Cell*. 2019;177(6):1600–1618. https://www.ncbi.nlm.nih.gov/pmc/articles/PMC6993574

Yap CX, AK Henders, GA Alvares, et al. Autism-related dietary preferences mediate autism-gut microbiome associations. *Cell*. 2021;184(24):5916–5931. https://pubmed.ncbi.nlm.nih.gov/34767757

Chapter 9. Remembrances of Things Past Haunt the Present

"Out of life's school of war: What does not kill me makes me stronger." —Friedrich Nietszche, "Twilight of the Idols," Aphorism 8 https://en.wikipedia.org/wiki/What_does_not_kill_me_makes_me_stronger

Brodin P, MM Davis. Human immune system variation. *Nat Rev Immunol*. 2017;(17):21–29. https://www.ncbi.nlm.nih.gov/pmc/articles/PMC5328245

Chapter 10. Harnessing the Function of the Immune System

Monoclonal antibodies: https://whatisbiotechnology.org/index.php/exhibitions/milstein/monoclonals/The-making-of-monoclonal-antibodies

Monaco C, J Nanchahal, P Taylor, M Feldmann. Anti-TNF therapy: past, present and future. *Int Immunol*. 2015;(27):55–62. https://www.ncbi.nlm.nih.gov/pmc/articles/PMC4279876

Karl Landsteiner and ABO blood groups: https://embryo.asu.edu/pages/karl-landsteiner-1868-1943

Tumor immunotherapy: https://www.nature.com/articles/ni.3613 https://www.mskcc.org/timeline/car-t-timeline-progress

Lim WA, CH June. The principles of engineering immune cells to treat cancer. *Cell*. 2017;(168):724–740. https://www.ncbi.nlm.nih.gov/pmc/articles/PMC5553442

Chapter 11. Vaccination

Li X, C Mukandavire, ZM Cucunuba, et al. Estimating the health impact of vaccination against ten pathogens in 98 low-income and middle-income countries from 2000 to 2030: a modelling study. *Lancet*. 2021;397(10272):398–408. https://www.ncbi.nlm.nih.gov/pmc/articles/PMC7846814

Riedel S. Edward Jenner and the history of smallpox and vaccination. *Proc (Bayl Univ Med Cent)*. 2005;18(1):21–25. https://www.ncbi.nlm.nih.gov/pmc/articles/PMC1200696

mRNA vaccines: https://laskerfoundation.org/winners/modified-mrna-vaccines https://www.nytimes.com/2020/11/10/business/biontech-covid-vaccine.html

"No man is an island entire of itself; every man is a piece of the continent, a part of the main; . . . any man's death diminishes me, because I am involved in mankind."
—John Donne, "Devotions Upon Emergent Occasions"
https://web.cs.dal.ca/~johnston/poetry/island.html

Chapter 12. "To Follow Knowledge Like a Sinking Star"
"To follow knowledge like a sinking star,
Beyond the utmost bound of human thought."
—Alfred, Lord Tennyson, "Ulysses"
https://www.poetryfoundation.org/poems/45392/Ulysses

Mak TW. The T cell antigen receptor: "The Hunting of the Snark." *Eur J Immunol.* 2007;37(suppl 1):S83–S93. https://onlinelibrary.wiley.com/doi/10.1002/eji.200737443

Janeway CA Jr. Approaching the asymptote? Evolution and revolution in immunology. *Cold Spring Harb Symp Quant Biol.* 1989;(54):1–13. https://www.jimmunol.org/content/191/9/4475.long

Cambri G, MT Mira. Genetic susceptibility to leprosy—from classic immune-related candidate genes to hypothesis-free, whole genome approaches. *Front Immunol.* 2018;(9):1674. https://doi.org/10.3389/fimmu.2018.01674

The letter *f* following a page number denotes a figure.

acetate. *See* short-chain fatty acid

acetylsalicylic acid (aspirin), 135

adaptive immunity, 2f, 8, 13, 19, 169; cell-mediated, 37, 48; humoral, 37, 52, 53, 57

ADCC (antibody-dependent, cell-mediated cytotoxicity), 58f, 59–60, 66, 143

adjuvant, 75, 151f, 153–54, 155, 169

affinity maturation, 56–57, 65

AIDS (acquired immunodeficiency syndrome), 84, 85–86, 112, 156, 159

AIRE gene, 92

allele, 72, 169

allelic exclusion, 43

allergy, 86f, 95–101, 169; and anaphylaxis, 97, 101, 169; and eosinophils, 97, 172; and genetics, 99–100; and hives, 6, 62, 96, 97, 169; and IgE, 61–62, 96, 99; and mast cells, 61–62, 62f, 96–97, 175; and microbiota, 111–12, 115; and oral tolerance, 98, 99; peanut, 95–96, 99, 100; and T cells, 98–99, 101

Allison, James, 143–44

Alzheimer's disease, 7, 101, 102

anaphylaxis, 97, 101, 169

anergy, 73, 99, 169

ankylosing spondylitis, 137

antibiotic, 13, 106, 114, 117

antibody, 37, 52f, 53, 170; and affinity maturation, 56–57, 65; blocking, 136–38; in diagnostic tests, 132–33; and effector functions, 57–62, 58f, 62f; isotypes, 53–54, 175; monoclonal, 133–35; mucosal, 60–61; natural, 142; and placental transport, 60; structure, 38–39; vaccine-induced, 150–53, 154

antibody-dependent, cell-mediated cytotoxicity (ADCC), 58f, 59–60, 66, 143

antigen, 36, 170; endogenous, 69, 69f, 70; epitopes, 134, 151f, 152, 153–54, 172; exogenous, 69, 69f, 70; foreign, 37, 44, 47, 48, 74, 98; processing, 67, 69, 69f, 74, 171; recognition, 36, 37, 44, 48; self, 46, 86f, 90, 91, 92, 170. *See also* antigen presentation; B Cell receptor; T Cell receptor

antigen presentation, 66–70, 69f

antigen receptor, 36f, 36–37. *See also* B cell receptor; T cell receptor

antihistamine, 97

anti-inflammatory agent, 135–36. *See also* corticosteroid

antimicrobial peptide, 12, 12f, 19, 80, 108, 110, 170

antitoxin, 131

aspirin (acetylsalicylic acid), 135

asthma, 103, 111–12

atherosclerosis, 7, 101, 102

ATP (adenosine triphosphate), 31, 170. *See also* DAMP

autism, 115–16

autoantibody, 90–91, 92

autoimmunity, 47, 86f, 90–95, 170. *See also* negative selection

B7-1, 73, 74f. *See also* Signal 2

Bacille Calmette-Guérin (BCG), 128, 157

bacteriophage, 113, 170
Baltimore, David, 44
B cell (B lymphocyte), 35, 36, 49; activation, 50f, 50–51, 53, 80, 152; clonal
  contraction, 52f, 63; clonal expansion,
  52, 52f, 53; clonal selection, 50, 52f;
  development, 35–36, 43; memory, 52f,
  64, 87; negative selection, 46, 91, 142
B cell receptor (BCR), 36, 36f; allelic
  exclusion, 43; diversification, 40–43,
  54–57; isotype switching, 54–56, 55f, 61,
  64, 83; somatic hypermutation, 56–57,
  64, 81; somatic recombination, 41–44,
  42f, 45, 46, 54
BCG (Bacille Calmette-Guérin), 128, 157
BCR. See B cell receptor
Belkaid, Yasmine, 125–26
Beutler, Bruce, 27
blood group, 141–42
blood transfusion, 141–42
bone marrow, 35, 64, 123, 170
bone marrow transplantation, 140
BTK gene, 88. See also X-linked
  agammaglobulinemia
bursa of Fabricius, 36
butyrate. See short-chain fatty acid

Candida albicans, 113
CAR (chimeric antigen receptor) T cell,
  145–47, 146f, 149
carrier protein, 154
CD (cluster of differentiation) molecule,
  51, 170; CD4, 73, 79, 81–82, 85; CD8, 73,
  77–78, 81, 83; CD19, 51, 146f, 147; CD28,
  73, 74f; CD40, 50f, 51; CD40L, 50f, 51, 80
celiac disease, 98, 100, 102
cell-mediated immunity, 37, 48, 52, 92.
  See also T cell
Celsus, 20
CGD (chronic granulomatous disease), 15,
  88
checkpoint inhibition, 143–45, 144f
chemokine, 16, 17–18, 19, 20f, 170
chimeric antigen receptor (CAR) T cell,
  145–47, 146f, 149
chromatin protein, 110, 128f, 129, 161, 170
chronic granulomatous disease (CGD),
  15, 88

citrulline, 91
clone (of lymphocytes), 52f, 52–53, 78, 78f,
  79, 171, 176, 178
Clostridium (Clostridioides) difficile, 13, 117
cluster of differentiation (CD) molecule.
  See CD (cluster of differentiation)
  molecule
CMV (cytomegalovirus), 123, 126–27
Coley Award, 157
complement, 34, 58f, 58–59, 63, 91, 171,
  178
coronavirus disease 2019. See COVID-19
corticosteroid, 136, 138, 171
co-stimulatory molecule, 73–75, 77, 98,
  153. See also Signal 2
COVID-19 (coronavirus disease 2019), 8,
  57, 95; type 1 interferon, 17, 89; vaccine,
  154–57. See also SARS-CoV-2
Crohn's disease, 93
cross-match, 142
cross-presentation, 70. See also antigen
  presentation
CTL (cytotoxic T lymphocyte), 77–79, 78f,
  79f
CTLA4, 82, 83, 143–44, 145. See also
  checkpoint inhibition
cyclosporine, 136, 141. See also immunosuppression; transplantation
cytokine, 16, 20f, 171. See also interferon;
  interleukin; tumor necrosis factor α
cytokine storm, 6, 17, 147, 171
cytomegalovirus (CMV), 123, 126–27
cytotoxic T lymphocyte (CTL), 77–79,
  78f, 79f

DAMP (damage-associated molecular
  pattern), 34, 35, 171, 174, 178; autoimmune disease, 32, 74–75; infection, 31,
  33, 76, 81; T cell activation, 67, 73, 75, 81,
  153, 169; trauma, 31, 140
Davis, Mark, 122, 126
dendritic cell, 2f, 3, 49, 75–76, 94, 154, 171,
  175; antigen processing, 67, 70, 73, 77;
  microbiota, 108, 109; T cell activation,
  51, 68–69, 74f, 74–75, 77, 79, 81, 98,
  153; vaccination, 153, 155–56. See also
  antigen presentation
diabetes, 85, 101–2

diarrhea: allergic, 62, 97, 98; infectious, 6, 14, 32, 111, 117, 123; inflammatory, 92
dietary fiber, 11, 109–10, 117, 118, 170

eczema, 99
Edelman, Gerald, 38–39, 40
EGID (eosinophilic gastrointestinal), 97–98
eicosanoid, 16, 18, 20f, 22, 76, 135, 136, 172
endocytosis, 51, 67, 68, 69f, 172
endoplasmic reticulum, 69f, 70, 171, 172
endosome, 28, 172
endothelial cell, 14, 18–19, 21, 138, 172. *See also* inflammation
endotoxin. *See* lipopolysaccharide
eosinophil, 97, 172
eosinophilic gastrointestinal disease (EGID), 97–98
epinephrine, 97, 172
epithelium: barrier, 2, 11–12, 12f, 14, 15, 93, 172; IgA, 60–61; microbiota, 10, 12–13, 32, 105, 108; skin, 10, 11, 12, 99, 109. *See also* mucosa
epitope, 134, 151f, 152, 153–54, 172
Epstein-Barr virus, 103
Escherichia coli, 32

Fcγ receptor, 58f, 59, 60
FcRn receptor, 60
fecal microbial transplantation (FMT), 116–17
Feldmann, Marc, 137
feral mice, 121–22
fever, 21, 77, 82, 86, 153
fibroblast, 14, 45, 173
filaggrin, 99, 100
flagellin, 25f, 28. *See also* PAMP
FMT (fecal microbial transplantation), 116–17
food additive, 118, 172
FOXP3 gene, 92
Freund's adjuvant, 75
Frost, Robert, 10, 13
fruit fly, 26, 27
fungus, 7, 26, 30, 37, 66, 104, 113

gastroenteritis, 22, 32, 112, 130
gastrointestinal tract, 13, 89, 92, 100, 173

genitourinary tract, 10, 173
germ-free mice, 105, 106, 111, 112, 115–16, 173
germinal center, 54, 56–57, 64
gluten, 98, 100, 102
glycan, 141, 142, 173, 176
goblet cell, 12f
Gordon, Jeffrey, 106, 107
gout, 138
graft-versus-host disease, 140
Gram, Hans Christian, 26
Gram-negative bacteria, 25, 27, 28, 109
Gram-positive bacteria, 27, 28
Graves' disease, 91

Hansen's disease. *See* leprosy
helper T cell (Th), 79–81; Tfh, 79f, 80–81; Th1, 79f, 80, 81, 89, 162; Th2, 79f, 80, 81, 99–100, 162; Th17, 79f, 80, 81, 109, 110, 125–26
hepatitis, 84, 132, 136, 153
herd immunity, 6, 158
HIV (human immunodeficiency virus), 85, 86, 112, 132, 156. *See also* AIDS
hives, 6, 62, 96, 97, 169
HLA (human leukocyte antigen). *See* MHC (major histocompatibility complex) molecule
Hoffmann, Jules, 26, 27
Honjo, Tasuku, 143, 144
human leukocyte antigen (HLA). *See* MHC (major histocompatibility complex) molecule
human milk oligosaccharides, 114
humoral immunity: adaptive, 37, 52, 53–55, 57, 60, 77, 152; innate, 63, 134; memory, 64. *See also* B cell; complement
hybridoma, 134–35, 173

IBD (inflammatory bowel disease), 92; genetics, 93–94; lifestyle factors, 103; microbiota, 94, 110; treatment, 117, 136, 137, 138; ultra-processed foods, 118
Ig. *See* immunoglobulin
IL. *See* interleukin
ILC (innate lymphoid cell), 33–34, 109, 173–74
immune amnesia, 87, 157

immune complex, 91
immunodeficiency, 85–90, 86f, 119, 146, 174
immunoglobulin (Ig), 37, 39, 53, 170; IgA,
    54, 57, 60–61, 110, 154; IgD, 62; IgE,
    54–55, 56, 61–62, 80, 96, 99–101; IgG,
    54, 56, 57–60, 63, 133; IgM, 54, 55, 57, 58,
    63, 133. *See also* antibody
immunosuppressant, 141, 174
immunosuppression, 140–41
infectious mononucleosis (mono), 103
inflammasome, 29, 33, 174
inflammation, 19–21, 20f; adhesion
    molecule, 18, 138; cell recruitment,
    18–19; fever, 21; local manifestations,
    20, 174; molecular mediators, 16–18;
    resolution, 21–22; sickness behavior, 21
inflammatory bowel disease. *See* IBD
influenza virus, 76, 81, 82, 126–27, 154
inhibitory receptor, 82, 83, 84. *See also*
    CTLA4; PD-1
innate immunity, 2f, 8; cells, 2–3, 14–19;
    concepts, 2–3, 8, 174; humoral, 34;
    innate lymphocyte, 33–34. *See also*
    complement; inflammation; innate
    lymphocyte; PAMP; pattern recogni-
    tion receptor; Toll-like receptor
innate lymphocyte, 33–34, 108, 173, 177
innate lymphoid cell (ILC), 33–34, 109,
    173–74
integrin, 138
interferon, 174; type 1, 17, 29, 76–77, 89,
    95, 174; type 2 (interferon γ), 33, 79f, 80,
    89, 111, 126, 174
interleukin (IL), 17, 174; IL-1β, 17, 29, 113,
    137, 138, 174; IL-4, 51, 79f, 80; IL-5, 79f,
    80; IL-6, 17, 29, 125, 129, 137, 138; IL-10,
    22, 79f; IL-13, 79f, 80; IL-17, 79f, 80, 138;
    IL-21, 51, 80
isotype switching, 54–56, 55f, 61, 64, 83.
    See also under B cell receptor

Janeway, Charles, 75, 153, 161
Jenner, Edward, 149
junctional proteins, 11–12, 12f

Karikó, Katalin, 155–57
Kitasato, Shibasaburo, 131
Köhler, Georges, 133–35

Landsteiner, Karl, 141
Lasker-DeBakey Medical Research Award,
    157
leprosy, 4–5, 66; genetics, 90, 93, 162;
    lepromatous, 5, 9, 66, 162; spectrum,
    5, 162; T helper response, 80, 162;
    tuberculoid, 5, 9, 80, 162
leukemia, 147
leukotriene, 18, 96. *See also* eicosanoid
lipopeptide, 25f, 28. *See also* PAMP
lipopolysaccharide (LPS), 25f, 25–26, 27,
    29, 32–33, 109, 153. *See also* PAMP
Listeria monocytogenes, 124, 127, 128
LPS. *See* lipopolysaccharide
lung cancer, 144
lupus. *See* systemic lupus erythematosus
lymph, 48, 49, 175
lymphatic vessel, 2f, 3, 48, 49, 77, 175
lymph node, 2f, 3, 48, 67, 108, 175; and
    B cells, 3, 47, 48, 53, 54, 63, 155; and
    dendritic cells, 3, 49, 75, 77, 155;
    and T cells, 47, 48, 73, 77, 79, 83, 155
lymphocyte, 2f, 8, 33, 175; adaptive, 3, 8,
    35, 37, 46–47, 48, 112, 119, 141; innate,
    33, 108. *See also* B cell; innate lympho-
    cyte; innate lymphoid cell; natural
    killer (NK) cell; T cell
lymphoid organ, 47, 48, 66. *See also* lymph
    node; spleen
lymphoma, 139, 147
lysosome, 15, 175

macrophage, 3, 14–16, 59, 80, 109, 128–29,
    175
Maini, Ravinder, 137
major histocompatibility complex (MHC)
    molecule. *See* MHC (major histocom-
    patibility complex) molecule
malaria, 112, 124, 159
malnutrition, 85
MAMP (microbe-associated molecular
    pattern), 25. *See also* PAMP
Masopust, David, 121
mast cell, 3, 61–62, 62f, 80, 96–97, 99,
    175
measles, 86–87, 150, 157, 158
Medzhitov, Ruslan, 105
melanoma, 144

memory, 37–38, 127; adaptive, 64, 83, 87, 109, 121, 130; innate, 128–29; microbiota-specific, 112; vaccine-induced, 150–52. *See also under* B cell; plasma cell; T cell; trained immunity

Mendelian susceptibility to mycobacterial disease (MSMD), 89

meta-organism, 11

Metchnikoff, Elie, 15–16

MHC (major histocompatibility complex) molecule, 51, 71, 72, 77; in antigen processing, 69f, 69–70; class I, 68f, 69–70, 77; class II, 67–68, 68f, 79, 100, 102; diversity, 72–73, 90; in transplantation, 139–41. *See also* antigen presentation; transplantation

microbe-associated molecular pattern (MAMP), 25. *See also* PAMP

microbiota, 7, 10–13, 104–5, 107, 176; allergic disease, 111–12; autism, 115–16; colonization resistance, 13; epithelial repair, 105–6; IBD, 92, 93–94; immunotherapy, 145; interactions with immune system, 32–33, 60–61, 81, 108f, 108–10; manipulation, 116–18; memory lymphocytes, 112; metabolites, 109–10; mycobiome, 113, 176; obesity, 106; variation, 113–15; virome, 113, 179

microglia, 21, 101, 176

micronutrient, 119, 176

Milstein, César, 133–35, 137

Moderna, 155, 156, 157

mono (infectious mononucleosis), 103

monocyte, 3, 19, 94, 175, 176

Montague, Lady Mary Wortley, 149

MSMD (Mendelian susceptibility to mycobacterial disease), 89

mucosa, 7, 11, 176

mucus, 12, 12f, 61, 62, 80, 110, 176

multiple sclerosis, 92, 94, 103

myasthenia gravis, 90

Mycobacterium bovis. See Bacille Calmette-Guérin

Mycobacterium leprae, 5, 9, 66, 80, 162

Mycobacterium tuberculosis, 6, 86–87, 89, 157

mycobiome, 113, 176

myelin, 92, 103, 176. *See also* multiple sclerosis

myeloma, 38–39, 134, 173, 176

natural killer (NK) cell, 33, 58f, 59, 129, 177

negative selection, 46–47, 71, 74, 91, 92, 139, 142

nerve cell (neuron), 14, 20f, 21, 92, 101, 176

neutralization, 57–58, 58f, 176

neutrophil, 2f, 18–19, 20f, 22, 59, 80, 177. *See also* inflammation

NK (natural killer) cell, 33, 58f, 59, 129, 177

Nobel Prize, 16, 27, 40, 44, 76, 131, 135, 144

NSAID (nonsteroidal anti-inflammatory drug), 135, 136

Nüsslein-Volhard, Christiane, 26, 27

obesity, 7, 101–2, 118

Oettinger, Marjorie, 45–46

opsonization, 57, 58f, 59, 60, 177

oral tolerance, 98, 99, 100–101

organelle, 14–15, 177

PAMP (pathogen-associated molecular pattern), 25, 25f, 27–28, 75, 109, 153, 177

Paneth cell, 12, 12f

parasite, 62, 80, 81, 112, 123

Pasteur, Louis, 149–50

pattern recognition receptor (PRR), 25, 27–31, 30f, 98, 153, 178. *See also* inflammasome; Toll-like receptor

PD-1, 82, 83, 143–44, 144f, 145

peanut allergy, 95–96, 99, 100–101, 130

Pfizer/BioNTech, 155–56

phagocytosis, 14, 15, 16, 31, 34, 63, 101, 177

phagosome, 14, 28, 177

Phipps, James, 149

plasma cell, 52f, 53, 60, 134; long-lived, 64, 84, 87, 130, 150–51, 151f, 152; short-lived, 54, 63

plasma membrane, 14, 27, 66, 146, 178

polio, 148, 159

poly-immunoglobulin receptor, 60

polymorphism, 90, 93, 100, 139, 162, 178

Porter, Rodney, 38–39, 40

positive selection, 71

postbiotic, 117, 118, 178
prebiotic, 117, 118, 178
pregnancy, 60, 125, 132, 142
probiotic, 117, 118, 178
programmed cell death, 46, 63, 82, 98
propionate. *See* short-chain fatty acid
prostaglandin, 18, 96, 135. *See also* eicosanoid
proteasome, 69f, 70, 178
protozoan, 28, 30, 66, 124, 178
PRR. *See* pattern recognition receptor
psoriasis, 137, 138
pus, 20

RAG (recombination activating gene), 46
Rakoff-Nahoum, Seth, 105, 106, 108
RBC (red blood cell), 134, 141, 142
reactive radicals (of oxygen or nitrogen), 14–15, 29. *See also* chronic granuloma- tous disease
receptor editing, 47. *See also* negative selection
recombination activating gene (RAG), 46
red blood cell (RBC), 134, 141, 142
regulatory T cell (Treg), 79f, 81, 82, 93, 99, 101, 109, 110
renal cancer, 144
respiratory tract, 10, 76, 126, 178
Rh antigen, 141, 142
rheumatoid arthritis, 32, 91, 94, 135, 137–38
Rossi, Derrick, 156
rotavirus, 6, 153, 154, 159

Sahin, Ugur, 156, 157
Salmonella enterica, 32, 37, 110, 125
SARS-CoV-2 (severe acute respiratory syndrome coronavirus-2), 6, 17, 57, 89, 132, 152, 154–56
SCFA (short-chain fatty acid), 109–10, 112, 117, 178
Schatz, David, 44–46
SCID (severe combined immunodeficiency), 89, 119
segmented filamentous bacteria, 109
self-nonself discrimination, 24, 25, 46, 47
sepsis, 86, 124, 178
severe acute respiratory syndrome coronavirus-2. *See* SARS-CoV-2

severe combined immunodeficiency (SCID), 89, 119
short-chain fatty acid (SCFA), 109–10, 112, 117, 178
sickness behavior, 21
Signal 1, 73, 74f, 74–75, 77, 146, 146f
Signal 2, 73, 74f, 74–75, 77, 98, 146, 146f
Signal 3, 74f, 75, 81
skin, 5, 11, 20, 99, 114, 137
SLE (systemic lupus erythematosus), 6, 7, 32, 91, 93, 135
smallpox, 148, 149, 159
somatic hypermutation, 56–57, 64, 81
somatic recombination, 41–44, 42f, 45, 46, 54
specialized pro-resolving mediator (SPM), 22, 82
spike protein, 57, 152–53, 155. *See also* SARS-CoV-2
spleen, 47, 48, 53, 54, 134
SPM (specialized pro-resolving mediator), 22, 82
Staphylococcus aureus, 37, 49, 52–53, 80
Steinman, Ralph, 27, 75–76
stem cell, 12, 12f, 156, 179
Streptococcus pneumoniae, 80
Streptococcus pyogenes, 132
stroma, 14, 179
switch region, 54, 55f
synaptic pruning, 101
systemic lupus erythematosus (SLE), 6, 7, 32, 91, 93, 135

tacrolimus, 136, 141. *See also* immunosuppression
TB (tuberculosis), 6, 86–87, 157, 159. *See also* Mycobacterium tuberculosis
T cell (T lymphocyte), 35, 66; activation, 73–75, 74f, 78f; CD4, 73, 81, 82, 85; CD8, 73, 78, 81, 82; cytotoxic (CTL), 77–79, 78f, 79f; exhaustion, 83–84; helper, 79–81, 79f, 109; memory, 78f, 83, 87, 109, 121, 122; negative selection, 46, 71, 74, 91, 92, 139; positive selection, 71; regulatory, 79f, 81, 93, 99, 101, 109; vaccination, 150, 151f, 152
T cell receptor (TCR), 36, 36f, 39, 160
T follicular helper cell (Tfh), 79f, 80

Th. *See* helper T cell
thymus, 35, 47, 71, 74, 92, 179
TLR (Toll-like receptor), 27–28, 32, 105–6, 109, 179
TNFα (tumor necrosis factor α), 17, 29, 96, 137, 138
tolerance, 98, 99, 100, 141, 179. *See also* oral tolerance
Toll gene, 26–27
Toll-like receptor (TLR), 27–28, 32, 105–6, 109, 179
Tonegawa, Susumu, 40, 41, 44
trained immunity, 128f, 128–29, 157, 179
transfusion reaction, 142
transplantation, 139–41
Treg. *See* regulatory T cell
tryptophan, 110
tuberculosis (TB), 6, 86–87, 157, 159. *See also Mycobacterium tuberculosis*
tumor immunotherapy, 143–47, 144f, 146f
tumor necrosis factor α (TNFα), 17, 29, 96, 137, 138
Türeci, Özlem, 156, 157
Turnbaugh, Peter, 106
twins, 72, 122–23, 140
typhoid, 80

ulcerative colitis, 93, 113
ultra-processed foods, 118

vaccination, 8, 148; adjuvant, 75, 151f, 153–54, 155, 169; efficacy, 148; herd immunity, 6, 158; history, 149–50; memory lymphocytes, 150–52, 151f; mRNA vaccine, 154–57; mucosal, 154; principles, 150–54, 151f; nonspecific protection, 157–58; T-dependent B cell activation, 152–54
Vane, John, 135
van Leeuwenhoek, Antonie, 104
variolation, 149, 150. *See also* vaccination
VDJ recombination. *See* somatic recombination
venom, 62, 96
virome, 113, 179
vitamins, 11, 13, 119, 176
von Behring, Emil, 131

weep and sweep mechanism, 62, 80
Weissman, Drew, 156, 157
Wieschaus, Eric, 27
William B. Coley Award, 157
worm, 30, 96, 122, 129

X chromosome, 94–95
X-linked agammaglobulinemia (XLA), 88–89
X-ray, 6, 87

Yersinia pseudotuberculosis, 125